James J. Buckley

Fuzzy Probability and Statistics

T0189166

Studies in Fuzziness and Soft Computing, Volume 196

Editor-in-chief
Prof. Janusz Kacprzyk
Systems Research Institute
Polish Academy of Sciences
ul. Newelska 6
01-447 Warsaw
Poland
E-mail: kacprzyk@ibspan.waw.pl

Further volumes of this series
can be found on our homepage:
springer.com

Vol. 180. Wesley Chu,
Tsau Young Lin (Eds.)
Foundations and Advances in Data Mining,
2005
ISBN 3-540-25057-3

Vol. 181. Nadia Nedjah,
Luiza de Macedo Mourelle
Fuzzy Systems Engineering, 2005
ISBN 3-540-25322-X

Vol. 182. John N. Mordeson,
Kiran R. Bhutani, Azriel Rosenfeld
Fuzzy Group Theory, 2005
ISBN 3-540-25072-7

Vol. 183. Larry Bull, Tim Kovacs (Eds.)
Foundations of Learning Classifier Systems,
2005
ISBN 3-540-25073-5

Vol. 184. Barry G. Silverman, Ashlesha Jain,
Ajita Ichalkaranje, Lakhmi C. Jain (Eds.)
*Intelligent Paradigms for Healthcare
Enterprises*, 2005
ISBN 3-540-22903-5

Vol. 185. Spiros Sirmakessis (Ed.)
Knowledge Mining, 2005
ISBN 3-540-25070-0

Vol. 186. Radim Bělohlávek, Vilém
Vychodil
Fuzzy Equational Logic, 2005
ISBN 3-540-26254-7

Vol. 187. Zhong Li, Wolfgang A. Halang,
Guanrong Chen (Eds.)
*Integration of Fuzzy Logic and Chaos
Theory*, 2006
ISBN 3-540-26899-5

Vol. 188. James J. Buckley, Leonard J.
Jowers
Simulating Continuous Fuzzy Systems, 2006
ISBN 3-540-28455-9

Vol. 189. Hans Bandemer
Mathematics of Uncertainty, 2006
ISBN 3-540-28457-5

Vol. 190. Ying-ping Chen
*Extending the Scalability of Linkage
Learning Genetic Algorithms*, 2006
ISBN 3-540-28459-1

Vol. 191. Martin V. Butz
*Rule-Based Evolutionary Online Learning
Systems*, 2006
ISBN 3-540-25379-3

Vol. 192. Jose A. Lozano, Pedro Larrañaga,
Iñaki Inza, Endika Bengoetxea (Eds.)
Towards a New Evolutionary Computation,
2006
ISBN 3-540-29006-0

Vol. 193. Ingo Glöckner
Fuzzy Quantifiers: A Computational Theory,
2006
ISBN 3-540-29634-4

Vol. 194. Dawn E. Holmes, Lakhmi C. Jain
(Eds.)
Innovations in Machince Learning, 2006
ISBN 3-540-30609-9

Vol. 195. Zongmin Ma
*Fuzzy Database Modeling of Imprecise and
Uncertain Engineering Information*, 2006
ISBN 3-540-30675-7

Vol. 196. James J. Buckley
Fuzzy Probability and Statistics, 2006
ISBN 3-540-30841-5

James J. Buckley

Fuzzy Probability
and Statistics

 Springer

Professor James J. Buckley
Department of Mathematics
University of Alabama at Birmingham
35294-1170 Birmingham, AL
USA
E-mail: buckley@math.uab.edu

ISSN print edition: 1434-9922
ISSN electronic edition: 1860-0808
ISBN-13 978-3-642-06809-6   ISBN-13 978-3-540-33190-2

Springer is a part of Springer Science+Business Media
springer.com
© Springer-Verlag Berlin Heidelberg 2006
Softcover reprint of the hardcover 1st edition 2006

To Julianne and Helen

Contents

1 Introduction   **1**
   1.1   Introduction   1
   1.2   Notation   3
   1.3   Previous Research   4
   1.4   Figures   4
   1.5   Maple/Solver Commands   5
   1.6   References   5

2 Fuzzy Sets   **7**
   2.1   Introduction   7
   2.2   Fuzzy Sets   7
       2.2.1   Fuzzy Numbers   8
       2.2.2   Alpha–Cuts   10
       2.2.3   Inequalities   10
       2.2.4   Discrete Fuzzy Sets   11
   2.3   Fuzzy Arithmetic   11
       2.3.1   Extension Principle   11
       2.3.2   Interval Arithmetic   12
       2.3.3   Fuzzy Arithmetic   13
   2.4   Fuzzy Functions   13
       2.4.1   Extension Principle   14
       2.4.2   Alpha–Cuts and Interval Arithmetic   15
       2.4.3   Differences   16
   2.5   Ordering Fuzzy Numbers   17
   2.6   References   18

3 Fuzzy Probability Theory   **21**
   3.1   Introduction   21
   3.2   Fuzzy Probabilities
       from Confidence Intervals   21
   3.3   Fuzzy Probabilities from Expert Opinion   23
   3.4   Restricted Fuzzy Arithmetic   24
       3.4.1   Probabilities   24

3.4.2   Restricted Arithmetic: General   25
3.4.3   Computing Fuzzy Probabilities   26
3.5   Fuzzy Probability   32
3.6   Fuzzy Conditional Probability   36
3.7   Fuzzy Independence   38
3.8   Fuzzy Bayes' Formula   40
3.9   Applications   41
3.9.1   Blood Types   41
3.9.2   Resistance to Surveys   43
3.9.3   Testing for HIV   43
3.9.4   Color Blindness   45
3.9.5   Fuzzy Bayes   47
3.10   References   48

4   Discrete Fuzzy Random Variables   51
4.1   Introduction   51
4.2   Fuzzy Binomial   51
4.3   Fuzzy Poisson   54
4.4   Applications   57
4.4.1   Fuzzy Poisson Approximating Fuzzy Binomial   57
4.4.2   Overbooking   58
4.4.3   Rapid Response Team   59
4.5   References   60

5   Continuous Fuzzy Random Variables   61
5.1   Introduction   61
5.2   Fuzzy Uniform   61
5.3   Fuzzy Normal   63
5.4   Fuzzy Negative Exponential   65
5.5   Applications   67
5.5.1   Fuzzy Uniform   67
5.5.2   Fuzzy Normal Approximation to Fuzzy Binomial . . .   67
5.5.3   Fuzzy Normal Approximation to Fuzzy Poisson   70
5.5.4   Fuzzy Normal   72
5.5.5   Fuzzy Negative Exponential   73
5.6   References   74

6   Estimate μ, Variance Known   75
6.1   Introduction   75
6.2   Fuzzy Estimation   75
6.3   Fuzzy Estimator of μ   76
6.4   References   79

7 Estimate μ, Variance Unknown   **81**
7.1 Fuzzy Estimator of μ 81
7.2 References 83

8 Estimate p, Binomial Population   **85**
8.1 Fuzzy Estimator of p 85
8.2 References 87

9 Estimate σ^2 from a Normal Population   **89**
9.1 Introduction 89
9.2 Biased Fuzzy Estimator 89
9.3 Unbiased Fuzzy Estimator 90
9.4 References 94

10 Fuzzy Arrival/Service Rates   **95**
10.1 Introduction 95
10.2 Fuzzy Arrival Rate 95
10.3 Fuzzy Service Rate 97
10.4 References 99

11 Fuzzy Uniform   **101**
11.1 Introduction 101
11.2 Fuzzy Estimators 101
    11.2.1 Details 101
11.3 References 105

12 Fuzzy Max Entropy Principle   **107**
12.1 Introduction 107
12.2 Maximum Entropy Principle 107
    12.2.1 Discrete Probability Distributions 107
    12.2.2 Continuous Probability Distributions 109
12.3 Imprecise Side-Conditions 111
    12.3.1 Discrete Probability Distributions 111
    12.3.2 Continuous Probability Distributions 112
12.4 Summary and Conclusions 113
12.5 References 114

13 Max Entropy: Crisp Discrete Solutions   **115**
13.1 Introduction 115
13.2 Max Entropy: Discrete Distributions 115
13.3 Max Entropy: Imprecise Side-Conditions 116
13.4 Summary and Conclusions 123
13.5 References 123

14 Max Entropy: Crisp Continuous Solutions   **125**
  14.1 Introduction 125
  14.2 Max Entropy: Probability Densities 126
  14.3 Max Entropy: Imprecise Side-Conditions 127
  14.4 $E = [0, M]$ 127
  14.5 $E = [0, \infty)$ 135
  14.6 $E = (-\infty, \infty)$ 139
  14.7 Summary and Conclusions 140
  14.8 References 140

15 Tests on μ, Variance Known   **143**
  15.1 Introduction 143
  15.2 Non-Fuzzy Case 143
  15.3 Fuzzy Case 144
  15.4 One-Sided Tests 148
  15.5 References 149

16 Tests on μ, Variance Unknown   **151**
  16.1 Introduction 151
  16.2 Crisp Case 151
  16.3 Fuzzy Model 152
      16.3.1 $\overline{T}[\alpha]$ for Non-Positive Intervals 153
  16.4 References 157

17 Tests on p for a Binomial Population   **159**
  17.1 Introduction 159
  17.2 Non-Fuzzy Test 159
  17.3 Fuzzy Test 160
  17.4 References 162

18 Tests on σ^2, Normal Population   **163**
  18.1 Introduction 163
  18.2 Crisp Hypothesis Test 163
  18.3 Fuzzy Hypothesis Test 164
  18.4 References 166

19 Fuzzy Correlation   **167**
  19.1 Introduction 167
  19.2 Crisp Results 167
  19.3 Fuzzy Theory 168
  19.4 References 170

20 Estimation in Simple Linear Regression   **171**
    20.1 Introduction 171
    20.2 Fuzzy Estimators 172
    20.3 References 175

21 Fuzzy Prediction in Linear Regression   **177**
    21.1 Prediction 177
    21.2 References 179

22 Hypothesis Testing in Regression   **181**
    22.1 Introduction 181
    22.2 Tests on a 181
    22.3 Tests on b 183
    22.4 References 185

23 Estimation in Multiple Regression   **187**
    23.1 Introduction 187
    23.2 Fuzzy Estimators 188
    23.3 References 192

24 Fuzzy Prediction in Regression   **193**
    24.1 Prediction 193
    24.2 References 195

25 Hypothesis Testing in Regression   **197**
    25.1 Introduction 197
    25.2 Tests on b 197
    25.3 Tests on c 199
    25.4 References 201

26 Fuzzy One-Way ANOVA   **203**
    26.1 Introduction 203
    26.2 Crisp Hypothesis Test 203
    26.3 Fuzzy Hypothesis Test 204
    26.4 References 207

27 Fuzzy Two-Way ANOVA   **209**
    27.1 Introduction 209
    27.2 Crisp Hypothesis Tests 209
    27.3 Fuzzy Hypothesis Tests 211
    27.4 References 217

28 Fuzzy Estimator for the Median   **219**
28.1 Introduction 219
28.2 Crisp Estimator for the Median 219
28.3 Fuzzy Estimator 220
28.4 Reference 221

29 Random Fuzzy Numbers   **223**
29.1 Introduction 223
29.2 Random Fuzzy Numbers 224
29.3 Tests for Randomness 225
     29.3.1 RNGenerator 227
     29.3.2 RNAnalysis 228
29.4 Monte Carlo Study 230
29.5 References 233

30 Selected Maple/Solver Commands   **235**
30.1 Introduction 235
30.2 SOLVER 235
     30.2.1 Example 13.3.1 235
     30.2.2 Example 13.3.2 236
     30.2.3 Example 13.3.3 236
     30.2.4 Example 13.3.4 236
     30.2.5 Problems 237
30.3 Maple 237
     30.3.1 Chapter 3 237
     30.3.2 Chapter 4 239
     30.3.3 Chapter 5 240
     30.3.4 Chapter 6 240
     30.3.5 Chapter 7 241
     30.3.6 Chapter 8 241
     30.3.7 Chapter 9 241
     30.3.8 Chapter 10 242
     30.3.9 Chapters 11–13 242
     30.3.10 Chapter 14 243
     30.3.11 Chapter 15 244
     30.3.12 Chapter 16 245
     30.3.13 Chapter 17 245
     30.3.14 Chapter 18 246
     30.3.15 Chapter 19 247
     30.3.16 Chapter 20 247
     30.3.17 Chapter 21 248
     30.3.18 Chapter 22 248
     30.3.19 Chapter 23 249
     30.3.20 Chapter 24 249
     30.3.21 Chapter 25 250

　　　30.3.22 Chapter 26 250
　　　30.3.23 Chapter 27 250
　　　30.3.24 Chapter 28 251
　　　30.3.25 Chapter 29 251
　　30.4 References 251

31 Summary and Future Research   253
　　31.1 Summary 253
　　31.2 Future Research 254
　　　31.2.1 Fuzzy Probability 254
　　　31.2.2 Unbiased Fuzzy Estimators 254
　　　31.2.3 Comparing Fuzzy Numbers 255
　　　31.2.4 No Decision Conclusion 255
　　　31.2.5 Fuzzy Uniform 255
　　　31.2.6 Interval Arithmetic 255
　　　31.2.7 Fuzzy Prediction 255
　　　31.2.8 Fuzzy ANOVA 256
　　　31.2.9 Nonparametric Statistics 256
　　　31.2.10 Randomness Tests Fuzzy Numbers 256
　　　31.2.11 Future 256
　　31.3 References 256

Index   257

List of Figures   265

List of Tables   269

Chapter 1

Introduction

1.1 Introduction

This book is written in the following divisions: (1) the introductory chapters consisting of Chapters 1 and 2; (2) introduction to fuzzy probability in Chapters 3-5; (3) introduction to fuzzy estimation in Chapters 6-11; (4) fuzzy/crisp estimators of probability density (mass) functions based on a fuzzy maximum entropy principle in Chapters 12-14; (5) introduction to fuzzy hypothesis testing in Chapters 15-18; (6) fuzzy correlation and regression in Chapters 19-25; (7) Chapters 26 and 27 are about a fuzzy ANOVA model; (8) a fuzzy estimator of the median in nonparametric statistics in Chapter 28; and (9) random fuzzy numbers with applications to Monte Carlo studies in Chapter 29.

First we need to be familiar with fuzzy sets. All you need to know about fuzzy sets for this book comprises Chapter 2. For a beginning introduction to fuzzy sets and fuzzy logic see [8]. One other item relating to fuzzy sets, needed in fuzzy hypothesis testing, is also in Chapter 2: how we will determine which of the following three possibilities is true $\overline{M} < \overline{N}$, $\overline{M} > \overline{N}$ or $\overline{M} \approx \overline{N}$, for two fuzzy numbers $\overline{M}$, $\overline{N}$.

The introduction to fuzzy probability in Chapters 3-5 is based on the book [1] and the reader is referred to that book for more information, especially applications. What is new here is: (1) using a nonlinear optimization program in Maple [13] to solve certain optimization problems in fuzzy probability, where previously we used a graphical method; and (2) a new algorithm, suitable for using only pencil and paper, for solving some restricted fuzzy arithmetic problems.

The introduction to fuzzy estimation is based on the book [3] and we refer the interested reader to that book for more about fuzzy estimators. The fuzzy estimators omitted from this book are those for $\mu_1 - \mu_2$, $p_1 - p_2$, σ_1/σ_2, etc. Fuzzy estimators for arrival and service rates is from [2] and [4]. The reader should see those book for applications in queuing networks.

James J. Buckley: *Fuzzy Probability and Statistics*, StudFuzz **196**, 1–6 (2006)
www.springerlink.com   © Springer-Verlag Berlin Heidelberg 2006

Also, fuzzy estimators for the uniform probability density can be found in [4], but the derivation of these fuzzy estimators is new to this book. The fuzzy uniform distribution was used for arrival/service rates in queuing models in [4].

The fuzzy/crisp probability density estimators based on a fuzzy maximum entropy principle are based on the papers [5],[6] and [7] and are new to this book. In Chapter 12 we obtain fuzzy results but in Chapters 13 and 14 we determine crisp discrete and crisp continuous probability densities.

The introduction to fuzzy hypothesis testing in Chapters 15-18 is based on the book [3] and the reader needs to consult that book for more fuzzy hypothesis testing. What we omitted are tests on $\mu_1 = \mu_2$, $p_1 = p_2$, $\sigma_1 = \sigma_2$, etc.

The chapters on fuzzy correlation and regression come from [3]. The results on the fuzzy ANOVA (Chapters 26 and 27) and a fuzzy estimator for the median (Chapter 28) are new and have not been published before.

The chapter on random fuzzy numbers (Chapter 29) is also new to this book and these results have not been previously published. Applications of crisp random numbers to Monte Carlo studies are well known and we also plan to use random fuzzy numbers in Monte Carlo studies. Our first use of random fuzzy numbers will be to get approximate solutions to fuzzy optimization problems whose solution is unknown or computationally very difficult. However, this becomes a rather large project and will probably be the topic of a future book.

Chapter 30 contains selected Maple/Solver ([11],[13],[20]) commands used in the book to solve optimization problems or to generate the figures. The final chapter has a summary and suggestions for future research.

All chapters can be read independently. This means that some material is repeated in a sequence of chapters. For example, in Chapters 15-18 on fuzzy hypothesis testing in each chapter we first review the crisp case, then fuzzify to obtain our fuzzy statistic which is then used to construct the fuzzy critical values and we finally present a numerical example. However, you should first know about fuzzy estimators (Chapters 6-11) before going on to fuzzy hypothesis testing.

A most important part of our models in fuzzy statistics is that we always start with a random sample producing crisp (non-fuzzy) data. Other authors discussing fuzzy statistics usually begin with fuzzy data. We assume we have a random sample giving real number data $x_1, x_2, ..., x_n$ which is then used to generate our fuzzy estimators. Using fuzzy estimators in hypothesis testing and regression obviously leads to fuzzy hypothesis testing and fuzzy regression.

Prerequisites, besides Chapter 2, are a basic knowledge of crisp elementary statistics. We will cover a lot of elementary statistics that can be found in Chapters 6-9 in [15]. We do not discuss contingency tables, or most of nonparametric statistics.

1.2 Notation

It is difficult, in a book with a lot of mathematics, to achieve a uniform notation without having to introduce many new specialized symbols. Our basic notation is presented in Chapter 2. What we have done is to have a uniform notation within each chapter. What this means is that we may use the letters "a" and "b" to represent a closed interval $[a, b]$ in one chapter but they could stand for parameters in a probability density in another chapter. We will have the following uniform notation throughout the book: (1) we place a "bar" over a letter to denote a fuzzy set ($\overline{A}$, $\overline{B}$, etc.), and all our fuzzy sets will be fuzzy subsets of the real numbers; and (2) an alpha-cut of a fuzzy set (Chapter 2) is always denoted by "α". Since we will be using α for alpha-cuts we need to change some standard notation in statistics: (1) we use β in confidence intervals; and (2) we will have γ as the significance level in hypothesis tests. So a $(1 - \beta)100\%$ confidence interval means a 95% confidence interval if $\beta = 0.05$. When a confidence interval switches to being an alpha-cut of a fuzzy number (see Chapter 6), we switch from β to α. Also a hypothesis test $H_0 : \mu = 0$ verses $H_1 : \mu \neq 0$ at $\gamma = 0.05$ means given that H_0 is true, the probability of landing in the critical region is 0.05.

All fuzzy arithmetic is performed using α-cuts and interval arithmetic and not by using the extension principle (Chapter 2). Fuzzy arithmetic is needed in fuzzy hypothesis testing, fuzzy prediction and Monte Carlo studies.

The term "crisp" means not fuzzy. A crisp set is a regular set and a crisp number is a real number. There is a potential problem with the symbol "$\leq$". It usually means "fuzzy subset" as $\overline{A} \leq \overline{B}$ stands for $\overline{A}$ is a fuzzy subset of $\overline{B}$ (defined in Chapter 2). However, also in Chapter 2, $\overline{A} \leq \overline{B}$ means that fuzzy set $\overline{A}$ is less than or equal to fuzzy set $\overline{B}$. The meaning of the symbol "$\leq$" should be clear from its use. Also, throughout the book $\overline{x}$ will be the mean of a random sample, not a fuzzy set, and we explicitly point this out when it first arises in the book and then usually not point it out again,

Let $N(\mu, \sigma^2)$ denote the normal distribution with mean μ and variance σ^2. Critical values for the normal will be written z_γ (z_β) for hypothesis testing (confidence intervals). We have $P(X \geq z_\gamma) = \gamma$. The binomial distribution is $b(n, p)$ where n is the number of independent trials and p is the probability of a "success". Critical values for the (Student's) t distribution are t_γ (t_β) so that $P(X \geq t_\gamma) = \gamma$. Critical values for the chi-square distribution are χ^2_γ (χ^2_β) so that $P(\chi^2 \geq \chi^2_\gamma) = \gamma$. We also use $\chi^2_{L,\beta/2}$ ($\chi^2_{R,\beta/2}$) where $P(\chi^2 \leq \chi^2_{L,\beta/2}) = \beta/2$ ($P(\chi^2 \geq \chi^2_{R,\beta/2}) = \beta/2$). Critical values for the F distribution are F_γ (F_β) where $P(X \geq F_\gamma) = \gamma$. The degrees of freedom associated with the t (χ^2, F) will all be stated when they are used and will usually not show up as subscripts in the symbol t (χ^2, F).

1.3   Previous Research

Our results on fuzzy probability falls in the intersection of the areas of imprecise probabilities, interval valued probabilities and fuzzy probabilities. See the references in Chapter 1 of [1] for a list of papers in these areas. The imprecise probability researchers have their own web site [19] which has links to basic papers in their area and to conferences on imprecise probabilities. The journal "Fuzzy Sets and Systems" recently (volume 154, September 2005) had a section devoted to fuzzy probabilities. Of course, you can always put "imprecise probability", "interval probability" or "fuzzy probability" into your favorite search engine.

Different from those papers on imprecise probabilities, which employ second order probabilities, possibilities, upper/lower probabilities, etc., we are using fuzzy numbers to model uncertainty in some of the probabilities, but we are not employing standard fuzzy arithmetic to combine the uncertainties. We could use crisp intervals to express the uncertainties but we would not be using standard interval arithmetic to combine the uncertainties. We do substitute fuzzy numbers for uncertain probabilities but we are not using fuzzy probability theory to propagate the uncertainty through the model. Our method is to use fuzzy numbers for imprecise probabilities and then through restricted fuzzy arithmetic calculate other fuzzy probabilities, expected values, variances, etc.

Statistical theory is based on probability theory. So fuzzy statistics can take many forms depending on what probability (imprecise, interval, fuzzy) theory you are using. A few key references to this relatively new area are ([9],[10],[12],[14],[16]-[18]) where you can find many more references. Also try "fuzzy statistics" in a search engine. The main difference with our method is that we always begin with crisp (non-fuzzy) data and other authors start with fuzzy data.

1.4   Figures

Some of the figures, graphs of certain fuzzy numbers, in the book are difficult to obtain so they were created using different methods. Many graphs were done first in Maple [13] and then exported to $LaTeX2_\epsilon$. We did these figures first in Maple because of the "implicitplot" command in Maple. Let us explain why this command was important in this book. Suppose $\overline{X}$ is a fuzzy estimator we want to graph. Usually in this book we determine $\overline{X}$ by first calculating its α-cuts. Let $\overline{X}[\alpha] = [x_1(\alpha), x_2(\alpha)]$. So we get $x = x_1(\alpha)$ describing the left side of the triangular shaped fuzzy number $\overline{X}$ and $x = x_2(\alpha)$ describes the right side. On a graph we would have the x-axis horizontal and the y-axis vertical. α is on the y-axis between zero and one. Substituting y for α we need to graph $x = x_i(y)$, for $i = 1, 2$. But this is backwards, we usually have y a function of x. The "implicitplot" command allows us to do the correct graph with x a function of y when we have $x = x_i(y)$. Figures 2.5,3.2,4.2,4.5,5.1,5.2,5.5,11.1-11.3,13.1,14.1,14.2,28.1 were constructed in LaTeX and all the rest in Maple first as described above.

1.5 Maple/Solver Commands

Some of the Maple commands we used to solve certain optimization problems and to create the figures are included in Chapter 30. Also, some of the Solver ([11],[20]) commands we used to solve the optimization problems in Chapter 13 are also included in Chapter 30.

1.6 References

1. J.J. Buckley: Fuzzy Probabilities: New Approach and Applications, Physica-Verlag, Heidelberg, Germany, 2003.

2. J.J. Buckley: Fuzzy Probabilities and Fuzzy Sets for Web Planning, Springer, Heidelberg, Germany, 2004.

3. J.J. Buckley: Fuzzy Statistics, Springer, Heidelberg, Germany, 2004.

4. J.J. Buckley: Simulating Fuzzy Systems, Springer, Heidelberg, Germany, 2005.

5. J.J. Buckley: Maximum Entropy Principle with Imprecise Side-Conditions, Soft Computing, 9(2005)507-511.

6. J.J. Buckley: Maximum Entropy Principle with Imprecise Side-Conditions II: Crisp Discrete Solutions, Soft Computing. To appear.

7. J.J. Buckley: Maximum Entropy Principle with Imprecise Side-Conditions III: Crisp Continuous Solutions, Soft Computing. To appear.

8. J.J. Buckley and E. Eslami: An Introduction to Fuzzy Logic and Fuzzy Sets, Springer-Verlag, Heidelberg, Germany, 2002.

9. T. Denoeux, M.-H. Masson and P.-A. Hebert: Nonparametric Rank-Based Statistics and Significance Tests for Fuzzy Data, Fuzzy Sets and Systems, 153(2005)1-28.

10. P. Filzmoser and R. Viertl: Testing Hypotheses with Fuzzy Data: the Fuzzy p-Value, Metrika, 59(2004)21-29.

11. Frontline Systems (www.frontsys.com).

12. M. Holena: Fuzzy Hypotheses Testing in the Framework of Fuzzy Logic, Fuzzy Sets and Systems, 145(2004)229-252.

13. Maple 9, Waterloo Maple Inc., Waterloo, Canada.

14. C. Romer and A. Kandel: Statistical Tests for Fuzzy Data, Fuzzy Sets and Systems, 72(1995)1-26.

15. M.J. Triola: Elementary Statistics Using Excel, Second Edition, Addison-Wesley, N.Y., 2004.

16. R. Viertl: Statistical Methods for Non-Precise Data, CRC Press, Boca Raton, FL, 1996.

17. N. Watanabe: A Fuzzy Statistical Test of Fuzzy Hypotheses: Fuzzy Sets and Systems, 25(1993)167-178.

18. H.-C. Wu: Statistical Hypotheses Testing for Fuzzy Data, Information Sciences, 175(2005)30-56.

19. www.sipta.org.

20. www.solver.com.

Chapter 2

Fuzzy Sets

2.1 Introduction

In this chapter we have collected together the basic ideas from fuzzy sets
and fuzzy functions needed for the book. Any reader familiar with fuzzy
sets, fuzzy numbers, the extension principle, α-cuts, interval arithmetic, and
fuzzy functions may go on and have a look at Section 2.5. In Section 2.5
we present a method of ordering a finite set of fuzzy numbers from smallest
to largest to be employed in fuzzy hypothesis testing, and in Monte Carlo
studies (Chapter 29). Basically, given two fuzzy numbers $\overline{M}$ and $\overline{N}$, we
need a method of deciding which of the following three possibilities is true:
$\overline{M} < \overline{N}, \overline{M} \approx \overline{N}, \overline{M} > \overline{N}$. A good general reference for fuzzy sets and fuzzy
logic is [4] and [10].

Our notation specifying a fuzzy set is to place a "bar" over a letter. So
$\overline{A}, \overline{B}, \ldots, \overline{X}, \overline{Y}, \ldots, \overline{\alpha}, \overline{\beta}, \ldots$, will all denote fuzzy sets.

2.2 Fuzzy Sets

If Ω is some set, then a fuzzy subset $\overline{A}$ of Ω is defined by its membership
function, written $\overline{A}(x)$, which produces values in $[0, 1]$ for all x in Ω. So, $\overline{A}(x)$
is a function mapping Ω into $[0, 1]$. If $\overline{A}(x_0) = 1$, then we say x_0 belongs to
$\overline{A}$, if $\overline{A}(x_1) = 0$ we say x_1 does not belong to $\overline{A}$, and if $\overline{A}(x_2) = 0.6$ we say
the membership value of x_2 in $\overline{A}$ is 0.6. When $\overline{A}(x)$ is always equal to one or
zero we obtain a crisp (non–fuzzy) subset of Ω. For all fuzzy sets $\overline{B}, \overline{C}, \ldots$
we use $\overline{B}(x), \overline{C}(x), \ldots$ for the value of their membership function at x. Most
of the fuzzy sets we will be using will be fuzzy numbers .

The term "crisp" will mean not fuzzy. A crisp set is a regular set. A
crisp number is just a real number. A crisp matrix (vector) has real numbers
as its elements. A crisp function maps real numbers (or real vectors) into

James J. Buckley: *Fuzzy Probability and Statistics*, StudFuzz **196**, 7–19 (2006)
www.springerlink.com

real numbers. A crisp solution to a problem is a solution involving crisp sets, crisp numbers, crisp functions, etc.

2.2.1   Fuzzy Numbers

A general definition of fuzzy number may be found in ([4],[10]), however our fuzzy numbers will be almost always triangular (shaped), or trapezoidal (shaped), fuzzy numbers. A triangular fuzzy number $\overline{N}$ is defined by three numbers $a < b < c$ where the base of the triangle is the interval $[a, c]$ and its vertex is at $x = b$. Triangular fuzzy numbers will be written as $\overline{N} = (a/b/c)$. A triangular fuzzy number $\overline{N} = (1.2/2/2.4)$ is shown in Figure 2.1. We see that $\overline{N}(2) = 1$, $\overline{N}(1.6) = 0.5$, etc.

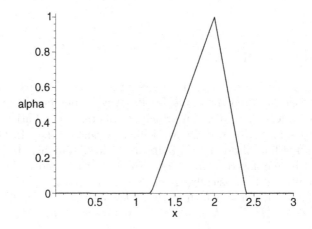

Figure 2.1: Triangular Fuzzy Number $\overline{N}$

A trapezoidal fuzzy number $\overline{M}$ is defined by four numbers $a < b < c < d$ where the base of the trapezoid is the interval $[a, d]$ and its top (where the membership equals one) is over $[b, c]$. We write $\overline{M} = (a/b, c/d)$ for trapezoidal fuzzy numbers. Figure 2.2 shows $\overline{M} = (1.2/2, 2.4/2.7)$.

A triangular shaped fuzzy number $\overline{P}$ is given in Figure 2.3. $\overline{P}$ is only partially specified by the three numbers 1.2, 2, 2.4 since the graph on $[1.2, 2]$, and $[2, 2.4]$, is not a straight line segment. To be a triangular shaped fuzzy number we require the graph to be continuous and: (1) monotonically increasing on $[1.2, 2]$; and (2) monotonically decreasing on $[2, 2.4]$. For triangular shaped fuzzy number $\overline{P}$ we use the notation $\overline{P} \approx (1.2/2/2.4)$ to show that it is partially defined by the three numbers 1.2, 2, and 2.4. If $\overline{P} \approx (1.2/2/2.4)$ we know its base is on the interval $[1.2, 2.4]$ with vertex (membership value one) at $x = 2$. Similarly we define trapezoidal shaped fuzzy number $\overline{Q} \approx (1.2/2, 2.4/2.7)$ whose base is $[1.2, 2.7]$ and top is over the interval $[2, 2.4]$. The graph of $\overline{Q}$ is similar to $\overline{M}$ in Figure 2.2 but it has continuous curves for its sides.

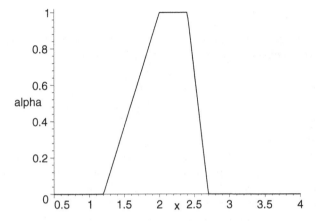

Figure 2.2: Trapezoidal Fuzzy Number $\overline{M}$

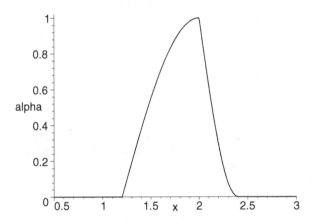

Figure 2.3: Triangular Shaped Fuzzy Number $\overline{P}$

Although we will be using triangular (shaped) and trapezoidal (shaped) fuzzy numbers throughout the book, many results can be extended to more general fuzzy numbers, but we shall be content to work with only these special fuzzy numbers.

We will be using fuzzy numbers in this book to describe uncertainty. For example, in Chapter 3 a fuzzy probability can be a triangular shaped fuzzy number, it could also be a trapezoidal shaped fuzzy number. In Chapters 4 and 5 parameters in probability density (mass) functions, like the mean in a normal probability density function, will be a triangular fuzzy number. Also, fuzzy estimators will be triangular shaped fuzzy numbers and fuzzy test statistics in fuzzy hypothesis testing are also triangular shaped fuzzy numbers.

2.2.2   Alpha–Cuts

Alpha–cuts are slices through a fuzzy set producing regular (non-fuzzy) sets. If $\overline{A}$ is a fuzzy subset of some set Ω, then an α–cut of $\overline{A}$, written $\overline{A}[\alpha]$ is defined as

$$\overline{A}[\alpha] = \{x \in \Omega | \overline{A}(x) \geq \alpha\} , \qquad (2.1)$$

for all α, $0 < \alpha \leq 1$. The $\alpha = 0$ cut, or $\overline{A}[0]$, must be defined separately.

Let $\overline{N}$ be the fuzzy number in Figure 2.1. Then $\overline{N}[0] = [1.2, 2.4]$. Notice that using equation (2.1) to define $\overline{N}[0]$ would give $\overline{N}[0] =$ all the real numbers. Similarly, $\overline{M}[0] = [1.2, 2.7]$ from Figure 2.2 and in Figure 2.3 $\overline{P}[0] = [1.2, 2.4]$. For any fuzzy set $\overline{A}$, $\overline{A}[0]$ is called the support, or base, of $\overline{A}$. Many authors call the support of a fuzzy number the open interval (a, b) like the support of $\overline{N}$ in Figure 2.1 would then be $(1.2, 2.4)$. However in this book we use the closed interval $[a, b]$ for the support (base) of the fuzzy number.

The core of a fuzzy number is the set of values where the membership value equals one. If $\overline{N} = (a/b/c)$, or $\overline{N} \approx (a/b/c)$, then the core of $\overline{N}$ is the single point b. However, if $\overline{M} = (a/b, c/d)$, or $\overline{M} \approx (a/b, c/d)$, then the core of $\overline{M} = [b, c]$.

For any fuzzy number $\overline{Q}$ we know that $\overline{Q}[\alpha]$ is a closed, bounded, interval for $0 \leq \alpha \leq 1$. We will write this as

$$\overline{Q}[\alpha] = [q_1(\alpha), q_2(\alpha)] , \qquad (2.2)$$

where $q_1(\alpha)$ ($q_2(\alpha)$) will be an increasing (decreasing) function of α with $q_1(1) \leq q_2(1)$. If $\overline{Q}$ is a triangular shaped or a trapezoidal shaped fuzzy number then: (1) $q_1(\alpha)$ will be a continuous, monotonically increasing function of α in $[0, 1]$; (2) $q_2(\alpha)$ will be a continuous, monotonically decreasing function of α, $0 \leq \alpha \leq 1$; and (3) $q_1(1) = q_2(1)$ ($q_1(1) < q_2(1)$ for trapezoids). We sometimes check monotone increasing (decreasing) by showing that $dq_1(\alpha)/d\alpha > 0$ ($dq_2(\alpha)/d\alpha < 0$) holds.

For the $\overline{N}$ in Figure 2.1 we obtain $\overline{N}[\alpha] = [n_1(\alpha), n_2(\alpha)]$, $n_1(\alpha) = 1.2 + 0.8\alpha$ and $n_2(\alpha) = 2.4 - 0.4\alpha$, $0 \leq \alpha \leq 1$. Similarly, $\overline{M}$ in Figure 2.2 has $\overline{M}[\alpha] = [m_1(\alpha), m_2(\alpha)]$, $m_1(\alpha) = 1.2 + 0.8\alpha$ and $m_2(\alpha) = 2.7 - 0.3\alpha$, $0 \leq \alpha \leq 1$. The equations for $n_i(\alpha)$ and $m_i(\alpha)$ are backwards. With the y–axis vertical and the x–axis horizontal the equation $n_1(\alpha) = 1.2 + 0.8\alpha$ means $x = 1.2 + 0.8y$, $0 \leq y \leq 1$. That is, the straight line segment from $(1.2, 0)$ to $(2, 1)$ in Figure 2.1 is given as x a function of y whereas it is usually stated as y a function of x. This is how it will be done for all α–cuts of fuzzy numbers.

2.2.3   Inequalities

Let $\overline{N} = (a/b/c)$. We write $\overline{N} \geq \delta$, δ some real number, if $a \geq \delta$, $\overline{N} > \delta$ when $a > \delta$, $\overline{N} \leq \delta$ for $c \leq \delta$ and $\overline{N} < \delta$ if $c < \delta$. We use the same notation for triangular shaped and trapezoidal (shaped) fuzzy numbers whose support is the interval $[a, c]$.

If $\overline{A}$ and $\overline{B}$ are two fuzzy subsets of a set Ω, then $\overline{A} \leq \overline{B}$ means $\overline{A}(x) \leq \overline{B}(x)$ for all x in Ω, or $\overline{A}$ is a fuzzy subset of $\overline{B}$. $\overline{A} < \overline{B}$ holds when $\overline{A}(x) < \overline{B}(x)$, for all x. There is a potential problem with the symbol $\leq$. In some places in the book , for example see Section 2.5 and in fuzzy hypothesis testing, $\overline{M} \leq \overline{N}$, for fuzzy numbers $\overline{M}$ and $\overline{N}$, means that $\overline{M}$ is less than or equal to $\overline{N}$. It should be clear on how we use "$\leq$" as to which meaning is correct.

2.2.4 Discrete Fuzzy Sets

Let $\overline{A}$ be a fuzzy subset of Ω. If $\overline{A}(x)$ is not zero only at a finite number of x values in Ω, then $\overline{A}$ is called a discrete fuzzy set. Suppose $\overline{A}(x)$ is not zero only at x_1, x_2, x_3 and x_4 in Ω. Then we write the fuzzy set as

$$\overline{A} = \{\frac{\mu_1}{x_1}, \cdots, \frac{\mu_4}{x_4}\}, \tag{2.3}$$

where the μ_i are the membership values. That is, $\overline{A}(x_i) = \mu_i$, $1 \leq i \leq 4$, and $\overline{A}(x) = 0$ otherwise. We can have discrete fuzzy subsets of any space Ω. Notice that α-cuts of discrete fuzzy sets of $\mathbb{R}$, the set of real numbers, do not produce closed, bounded, intervals.

2.3 Fuzzy Arithmetic

If $\overline{A}$ and $\overline{B}$ are two fuzzy numbers we will need to add, subtract, multiply and divide them. There are two basic methods of computing $\overline{A} + \overline{B}$, $\overline{A} - \overline{B}$, etc. which are: (1) extension principle; and (2) α–cuts and interval arithmetic.

2.3.1 Extension Principle

Let $\overline{A}$ and $\overline{B}$ be two fuzzy numbers. If $\overline{A} + \overline{B} = \overline{C}$, then the membership function for $\overline{C}$ is defined as

$$\overline{C}(z) = \sup_{x,y}\{\min(\overline{A}(x), \overline{B}(y))|x + y = z\} . \tag{2.4}$$

If we set $\overline{C} = \overline{A} - \overline{B}$, then

$$\overline{C}(z) = \sup_{x,y}\{\min(\overline{A}(x), \overline{B}(y))|x - y = z\} . \tag{2.5}$$

Similarly, $\overline{C} = \overline{A} \cdot \overline{B}$, then

$$\overline{C}(z) = \sup_{x,y}\{\min(\overline{A}(x), \overline{B}(y))|x \cdot y = z\} , \tag{2.6}$$

and if $\overline{C} = \overline{A}/\overline{B}$,

$$\overline{C}(z) = \sup_{x,y}\{\min(\overline{A}(x), \overline{B}(y))|x/y = z\} . \tag{2.7}$$

In all cases $\overline{C}$ is also a fuzzy number [10]. We assume that zero does not belong to the support of $\overline{B}$ in $\overline{C} = \overline{A}/\overline{B}$. If $\overline{A}$ and $\overline{B}$ are triangular (trapezoidal) fuzzy numbers then so are $\overline{A} + \overline{B}$ and $\overline{A} - \overline{B}$, but $\overline{A} \cdot \overline{B}$ and $\overline{A}/\overline{B}$ will be triangular (trapezoidal) shaped fuzzy numbers.

We should mention something about the operator "sup" in equations (2.4) – (2.7). If Ω is a set of real numbers bounded above (there is a M so that $x \leq M$, for all x in Ω), then $\sup(\Omega) =$ the least upper bound for Ω. If Ω has a maximum member, then $\sup(\Omega) = \max(\Omega)$. For example, if $\Omega = [0,1)$, $\sup(\Omega) = 1$ but if $\Omega = [0,1]$, then $\sup(\Omega) = \max(\Omega) = 1$. The dual operator to "sup" is "inf". If Ω is bounded below (there is a M so that $M \leq x$ for all $x \in \Omega$), then $\inf(\Omega) =$ the greatest lower bound. For example, for $\Omega = (0,1]$ $\inf(\Omega) = 0$ but if $\Omega = [0,1]$, then $\inf(\Omega) = \min(\Omega) = 0$.

Obviously, given $\overline{A}$ and $\overline{B}$, equations (2.4) – (2.7) appear quite complicated to compute $\overline{A} + \overline{B}$, $\overline{A} - \overline{B}$, etc. So, we now present an equivalent procedure based on α–cuts and interval arithmetic. First, we present the basics of interval arithmetic.

2.3.2   Interval Arithmetic

We only give a brief introduction to interval arithmetic. For more information the reader is referred to ([12],[13]). Let $[a_1, b_1]$ and $[a_2, b_2]$ be two closed, bounded, intervals of real numbers. If $*$ denotes addition, subtraction, multiplication, or division, then $[a_1, b_1] * [a_2, b_2] = [\alpha, \beta]$ where

$$[\alpha, \beta] = \{a * b | a_1 \leq a \leq b_1, a_2 \leq b \leq b_2\} . \tag{2.8}$$

If $*$ is division, we must assume that zero does not belong to $[a_2, b_2]$. We may simplify equation (2.8) as follows:

$$[a_1, b_1] + [a_2, b_2] = [a_1 + a_2, b_1 + b_2] , \tag{2.9}$$

$$[a_1, b_1] - [a_2, b_2] = [a_1 - b_2, b_1 - a_2] , \tag{2.10}$$

$$[a_1, b_1] / [a_2, b_2] = [a_1, b_1] \cdot \left[\frac{1}{b_2}, \frac{1}{a_2}\right] , \tag{2.11}$$

and

$$[a_1, b_1] \cdot [a_2, b_2] = [\alpha, \beta] , \tag{2.12}$$

where

$$\alpha = \min\{a_1 a_2, a_1 b_2, b_1 a_2, b_1 b_2\} , \tag{2.13}$$

$$\beta = \max\{a_1 a_2, a_1 b_2, b_1 a_2, b_1 b_2\} . \tag{2.14}$$

Multiplication and division may be further simplified if we know that $a_1 > 0$ and $b_2 < 0$, or $b_1 > 0$ and $b_2 < 0$, etc. For example, if $a_1 \geq 0$ and $a_2 \geq 0$, then

$$[a_1, b_1] \cdot [a_2, b_2] = [a_1 a_2, b_1 b_2] , \tag{2.15}$$

and if $b_1 < 0$ but $a_2 \geq 0$, we see that

$$[a_1, b_1] \cdot [a_2, b_2] = [a_1 b_2, a_2 b_1] . \tag{2.16}$$

Also, assuming $b_1 < 0$ and $b_2 < 0$ we get

$$[a_1, b_1] \cdot [a_2, b_2] = [b_1 b_2, a_1 a_2] , \tag{2.17}$$

but $a_1 \geq 0$, $b_2 < 0$ produces

$$[a_1, b_1] \cdot [a_2, b_2] = [a_2 b_1, b_2 a_1] . \tag{2.18}$$

2.3.3 Fuzzy Arithmetic

Again we have two fuzzy numbers $\overline{A}$ and $\overline{B}$. We know α–cuts are closed, bounded, intervals so let $\overline{A}[\alpha] = [a_1(\alpha), a_2(\alpha)]$, $\overline{B}[\alpha] = [b_1(\alpha), b_2(\alpha)]$. Then if $\overline{C} = \overline{A} + \overline{B}$ we have

$$\overline{C}[\alpha] = \overline{A}[\alpha] + \overline{B}[\alpha] . \tag{2.19}$$

We add the intervals using equation (2.9). Setting $\overline{C} = \overline{A} - \overline{B}$ we get

$$\overline{C}[\alpha] = \overline{A}[\alpha] - \overline{B}[\alpha] , \tag{2.20}$$

for all α in $[0, 1]$. Also

$$\overline{C}[\alpha] = \overline{A}[\alpha] \cdot \overline{B}[\alpha] , \tag{2.21}$$

for $\overline{C} = \overline{A} \cdot \overline{B}$ and

$$\overline{C}[\alpha] = \overline{A}[\alpha] / \overline{B}[\alpha] , \tag{2.22}$$

when $\overline{C} = \overline{A}/\overline{B}$, provided that zero does not belong to $\overline{B}[\alpha]$ for all α. This method is equivalent to the extension principle method of fuzzy arithmetic [10]. Obviously, this procedure, of α–cuts plus interval arithmetic, is more user (and computer) friendly.

Example 2.3.3.1

Let $\overline{A} = (-3/-2/-1)$ and $\overline{B} = (4/5/6)$. We determine $\overline{A} \cdot \overline{B}$ using α–cuts and interval arithmetic. We compute $\overline{A}[\alpha] = [-3 + \alpha, -1 - \alpha]$ and $\overline{B}[\alpha] = [4 + \alpha, 6 - \alpha]$. So, if $\overline{C} = \overline{A} \cdot \overline{B}$ we obtain $\overline{C}[\alpha] = [(\alpha - 3)(6 - \alpha), (-1 - \alpha)(4 + \alpha)]$, $0 \leq \alpha \leq 1$. The graph of $\overline{C}$ is shown in Figure 2.4.

2.4 Fuzzy Functions

In this book a fuzzy function is a mapping from fuzzy numbers into fuzzy numbers. We write $H(\overline{X}) = \overline{Z}$ for a fuzzy function with one independent variable $\overline{X}$. Usually $\overline{X}$ will be a triangular (trapezoidal) fuzzy number and

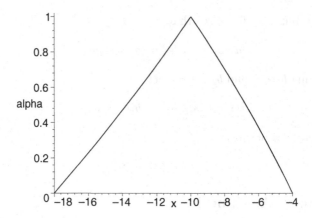

Figure 2.4: The Fuzzy Number $\overline{C} = \overline{A} \cdot \overline{B}$

then we usually obtain $\overline{Z}$ as a triangular (trapezoidal) shaped fuzzy number. For two independent variables we have $H(\overline{X}, \overline{Y}) = \overline{Z}$.

Where do these fuzzy functions come from? They are usually extensions of real–valued functions. Let $h : [a, b] \to \mathbb{R}$. This notation means $z = h(x)$ for x in $[a, b]$ and z a real number. One extends $h : [a, b] \to \mathbb{R}$ to $H(\overline{X}) = \overline{Z}$ in two ways: (1) the extension principle; or (2) using α–cuts and interval arithmetic.

2.4.1   Extension Principle

Any $h : [a, b] \to \mathbb{R}$ may be extended to $H(\overline{X}) = \overline{Z}$ as follows

$$\overline{Z}(z) = \sup_{x} \left\{ \, \overline{X}(x) \mid h(x) = z, \ a \leq x \leq b \, \right\} . \tag{2.23}$$

Equation (2.23) defines the membership function of $\overline{Z}$ for any triangular (trapezoidal) fuzzy number $\overline{X}$ in $[a, b]$.

If h is continuous, then we have a way to find α–cuts of $\overline{Z}$. Let $\overline{Z}[\alpha] = [z_1(\alpha), z_2(\alpha)]$. Then [6]

$$z_1(\alpha) = \min\{ \, h(x) \mid x \in \overline{X}[\alpha] \, \} , \tag{2.24}$$
$$z_2(\alpha) = \max\{ \, h(x) \mid x \in \overline{X}[\alpha] \, \} , \tag{2.25}$$

for $0 \leq \alpha \leq 1$.

If we have two independent variables, then let $z = h(x, y)$ for x in $[a_1, b_1]$, y in $[a_2, b_2]$. We extend h to $H(\overline{X}, \overline{Y}) = \overline{Z}$ as

$$\overline{Z}(z) = \sup_{x,y} \left\{ \min \left(\overline{X}(x), \overline{Y}(y) \right) \mid h(x, y) = z \, \right\} , \tag{2.26}$$

for $\overline{X}$ $(\overline{Y})$ a triangular or trapezoidal fuzzy number in $[a_1, b_1]$ $([a_2, b_2])$. For α–cuts of $\overline{Z}$, assuming h is continuous, we have

$$z_1(\alpha) \;=\; \min\{\, h(x, y) \mid x \in \overline{X}[\alpha],\; y \in \overline{Y}[\alpha] \,\} \,, \qquad (2.27)$$
$$z_2(\alpha) \;=\; \max\{\, h(x, y) \mid x \in \overline{X}[\alpha],\; y \in \overline{Y}[\alpha] \,\} \,, \qquad (2.28)$$

$0 \leq \alpha \leq 1$. We use equations (2.24) – (2.25) and (2.27) – (2.28) throughout this book.

Applications

Let $f(x_1, ..., x_n; \theta_1, ..., \theta_m)$ be a continuous function . Then

$$I[\alpha] = \{f(x_1, ..., x_n; \theta_1, ..., \theta_m)| \quad \mathbf{S} \quad \}, \qquad (2.29)$$

for $\alpha \in [0, 1]$ and $\mathbf{S}$ is the statement "$\theta_i \in \overline{\theta}_i[\alpha]$, $1 \leq i \leq m$", for fuzzy numbers $\overline{\theta}_i$, $1 \leq i \leq m$, defines an interval $I[\alpha]$. The endpoints of $I[\alpha]$ may be found as in equations (2.24),(2.25) and (2.27),(2.28). $I[\alpha]$ gives the α-cuts of $f(x_1, ..., x_n; \overline{\theta}_i, ..., \overline{\theta}_m)$.

This result will be used throughout the book.

2.4.2 Alpha–Cuts and Interval Arithmetic

All the functions we usually use in engineering and science have a computer algorithm which, using a finite number of additions, subtractions, multiplications and divisions, can evaluate the function to required accuracy [5]. Such functions can be extended, using α–cuts and interval arithmetic, to fuzzy functions. Let $h : [a, b] \to \mathbb{R}$ be such a function. Then its extension $H(\overline{X}) = \overline{Z}$, $\overline{X}$ in $[a, b]$ is done, via interval arithmetic, in computing $h(\overline{X}[\alpha]) = \overline{Z}[\alpha]$, α in $[0, 1]$. We input the interval $\overline{X}[\alpha]$, perform the arithmetic operations needed to evaluate h on this interval, and obtain the interval $\overline{Z}[\alpha]$. Then put these α–cuts together to obtain the value $\overline{Z}$. The extension to more independent variables is straightforward.

For example, consider the fuzzy function

$$\overline{Z} = H(\overline{X}) = \frac{\overline{A}\,\overline{X} + \overline{B}}{\overline{C}\,\overline{X} + \overline{D}} \,, \qquad (2.30)$$

for triangular fuzzy numbers $\overline{A}$, $\overline{B}$, $\overline{C}$, $\overline{D}$ and triangular fuzzy number $\overline{X}$ in $[0, 10]$. We assume that $\overline{C} \geq 0$, $\overline{D} > 0$ so that $\overline{C}\,\overline{X} + \overline{D} > 0$. This would be the extension of

$$h(x_1, x_2, x_3, x_4, x) = \frac{x_1 x + x_2}{x_3 x + x_4} \,. \qquad (2.31)$$

We would substitute the intervals $\overline{A}[\alpha]$ for x_1, $\overline{B}[\alpha]$ for x_2, $\overline{C}[\alpha]$ for x_3, $\overline{D}[\alpha]$ for x_4 and $\overline{X}[\alpha]$ for x, do interval arithmetic, to obtain interval $\overline{Z}[\alpha]$ for $\overline{Z}$.

Alternatively, the fuzzy function

$$\overline{Z} = H(\overline{X}) = \frac{2\overline{X} + 10}{3\overline{X} + 4} \; , \tag{2.32}$$

would be the extension of

$$h(x) = \frac{2x + 10}{3x + 4} \; . \tag{2.33}$$

2.4.3   Differences

Let $h : [a, b] \to \mathbb{R}$. Just for this subsection let us write $\overline{Z}^* = H(\overline{X})$ for the extension principle method of extending h to H for $\overline{X}$ in $[a, b]$. We denote $\overline{Z} = H(\overline{X})$ for the α–cut and interval arithmetic extension of h.

We know that $\overline{Z}$ can be different from $\overline{Z}^*$. But for basic fuzzy arithmetic in Section 2.3 the two methods give the same results. In the example below we show that for $h(x) = x(1 - x)$, x in $[0, 1]$, we can get $\overline{Z}^* \neq \overline{Z}$ for some $\overline{X}$ in $[0, 1]$. What is known ([6],[12]) is that for usual functions in science and engineering $\overline{Z}^* \leq \overline{Z}$. Otherwise, there is no known necessary and sufficient conditions on h so that $\overline{Z}^* = \overline{Z}$ for all $\overline{X}$ in $[a, b]$. See also [11].

There is nothing wrong in using α–cuts and interval arithmetic to evaluate fuzzy functions. Surely, it is user, and computer friendly. However, we should be aware that whenever we use α–cuts plus interval arithmetic to compute $\overline{Z} = H(\overline{X})$ we may be getting something larger than that obtained from the extension principle. The same results hold for functions of two or more independent variables.

Example 2.4.3.1

The example is the simple fuzzy expression

$$\overline{Z} = (1 - \overline{X}) \, \overline{X} \; , \tag{2.34}$$

for $\overline{X}$ a triangular fuzzy number in $[0, 1]$. Let $\overline{X}[\alpha] = [x_1(\alpha), x_2(\alpha)]$. Using interval arithmetic we obtain

$$z_1(\alpha) = (1 - x_2(\alpha))x_1(\alpha) \; , \tag{2.35}$$
$$z_2(\alpha) = (1 - x_1(\alpha))x_2(\alpha) \; , \tag{2.36}$$

for $\overline{Z}[\alpha] = [z_1(\alpha), z_2(\alpha)]$, α in $[0, 1]$.

The extension principle extends the regular equation $z = (1 - x)x$, $0 \leq x \leq 1$, to fuzzy numbers as follows

$$\overline{Z}^*(z) = \sup_x \left\{ \overline{X}(x) | (1 - x)x = z, \; 0 \leq x \leq 1 \right\} \; . \tag{2.37}$$

Let $\overline{Z}^*[\alpha] = [z_1^*(\alpha), z_2^*(\alpha)]$. Then

$$z_1^*(\alpha) = \min\{(1-x)x | x \in \overline{X}[\alpha]\}, \tag{2.38}$$

$$z_2^*(\alpha) = \max\{(1-x)x | x \in \overline{X}[\alpha]\}, \tag{2.39}$$

for all $0 \leq \alpha \leq 1$. Now let $\overline{X} = (0/0.25/0.5)$, then $x_1(\alpha) = 0.25\alpha$ and $x_2(\alpha) = 0.50 - 0.25\alpha$. Equations (2.35) and (2.36) give $\overline{Z}[0.50] = [5/64, 21/64]$ but equations (2.38) and (2.39) produce $\overline{Z}^*[0.50] = [7/64, 15/64]$. Therefore, $\overline{Z}^* \neq \overline{Z}$. We do know that if each fuzzy number appears only once in the fuzzy expression, the two methods produce the same results ([6],[12]). However, if a fuzzy number is used more than once, as in equation (2.34), the two procedures can give different results.

2.5 Ordering Fuzzy Numbers

Given a finite set of fuzzy numbers $\overline{A}_1, ..., \overline{A}_n$ we want to order them from smallest to largest. For a finite set of real numbers there is no problem in ordering them from smallest to largest. However, in the fuzzy case there is no universally accepted way to do this. There are probably more than 40 methods proposed in the literature of defining $\overline{M} \leq \overline{N}$, for two fuzzy numbers $\overline{M}$ and $\overline{N}$. Here the symbol $\leq$ means "less than or equal" and not "a fuzzy subset of". A few key references on this topic are ([1],[7]-[9],[14],[15]), where the interested reader can look up many of these methods and see their comparisons.

Here we will present only one procedure for ordering fuzzy numbers that we have used before ([2],[3]). But note that different definitions of $\leq$ between fuzzy numbers can give different ordering. We first define $<$ between two fuzzy numbers $\overline{M}$ and $\overline{N}$. Define

$$v(\overline{M} \leq \overline{N}) = max\{min(\overline{M}(x), \overline{N}(y)) | x \leq y\}, \tag{2.40}$$

which measures how much $\overline{M}$ is less than or equal to $\overline{N}$. We write $\overline{N} < \overline{M}$ if $v(\overline{N} \leq \overline{M}) = 1$ but $v(\overline{M} \leq \overline{N}) < \eta$, where η is some fixed fraction in $(0, 1]$. In this book we will usually use $\eta = 0.8$. Then $\overline{N} < \overline{M}$ if $v(\overline{N} \leq \overline{M}) = 1$ and $v(\overline{M} \leq \overline{N}) < 0.8$. We then define $\overline{M} \approx \overline{N}$ when both $\overline{N} < \overline{M}$ and $\overline{M} < \overline{N}$ are false. $\overline{M} \leq \overline{N}$ means $\overline{M} < \overline{N}$ or $\overline{M} \approx \overline{N}$. Now this $\approx$ may not be transitive. If $\overline{N} \approx \overline{M}$ and $\overline{M} \approx \overline{O}$ implies that $\overline{N} \approx \overline{O}$, then $\approx$ is transitive. However, it can happen that $\overline{N} \approx \overline{M}$ and $\overline{M} \approx \overline{O}$ but $\overline{N} < \overline{O}$ because $\overline{M}$ lies a little to the right of $\overline{N}$ and $\overline{O}$ lies a little to the right of $\overline{M}$ but $\overline{O}$ lies sufficiently far to the right of $\overline{N}$ that we obtain $\overline{N} < \overline{O}$.

But this ordering is still useful in partitioning the set of fuzzy numbers $\overline{A}_i$, $1 \leq i \leq n$, up into disjoint sets $H_1, ..., H_K$ where ([2],[3]): (1) given any $\overline{A}_i$ and $\overline{A}_j$ in H_k, $1 \leq k \leq K$, then $\overline{A}_i \approx \overline{A}_j$; and (2) given $\overline{A}_i \in H_i$ and $i < j$, there is a $\overline{A}_j \in H_j$ with $\overline{A}_i < \overline{A}_j$. We say a fuzzy number $\overline{A}_i$ is dominated if there is another fuzzy number $\overline{A}_j$ so that $\overline{A}_i < \overline{A}_j$. So H_K will

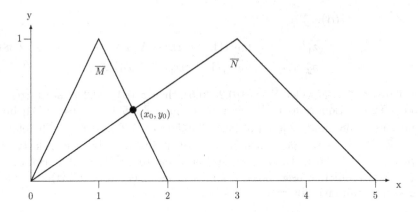

Figure 2.5: Determining $v(\overline{N} \leq \overline{M})$

be all the undominated $\overline{A}_i$. Now H_K is nonempty and if it does not contain all the fuzzy numbers we then define H_{K-1} to be all the undominated fuzzy numbers after we delete all those in H_K. We continue this way to the last set H_1. Then the highest ranked fuzzy numbers lie in H_K, the second highest ranked fuzzy numbers are in H_{K-1}, etc. This result is easily seen if you graph all the fuzzy numbers on the same axis then those in H_K will be clustered together farthest to the right, proceeding from the H_K cluster to the left the next cluster will be those in H_{K-1}, etc.

There is an easy way to determine if $\overline{M} < \overline{N}$, or $\overline{M} \approx \overline{N}$, for many fuzzy numbers. This will be all we need in fuzzy hypothesis testing and Monte Carlo studies. First, it is easy to see that if the core of $\overline{N}$ lies completely to the right of the core of $\overline{M}$, then $v(\overline{M} \leq \overline{N}) = 1$. Also, if the core of $\overline{M}$ and the core of $\overline{N}$ overlap, then $\overline{M} \approx \overline{N}$. Now assume that the core of $\overline{N}$ lies to the right of the core of $\overline{M}$, as shown in Figure 2.5 for triangular fuzzy numbers, and we wish to compute $v(\overline{N} \leq \overline{M})$. The value of this expression is simply y_0 in Figure 2.5. In general, for triangular (shaped), and trapezoidal (shaped), fuzzy numbers $v(\overline{N} \leq \overline{M})$ is the height of their intersection when the core of $\overline{N}$ lies to the right of the core of $\overline{M}$.

2.6   References

1. G. Bortolon and R. Degani: A Review of Some Methods for Ranking Fuzzy Subsets, Fuzzy Sets and Systems, 15(1985)1-19.

2. J.J. Buckley: Ranking Alternatives Using Fuzzy Numbers, Fuzzy Sets and Systems, 15(1985)21-31.

3. J.J. Buckley: Fuzzy Hierarchical Analysis, Fuzzy Sets and Systems, 17(1985)233-247.

4. J.J. Buckley and E. Eslami: Introduction to Fuzzy Logic and Fuzzy Sets, Physica-Verlag, Heidelberg, Germany, 2002.

5. J.J. Buckley and Y. Hayashi: Can Neural Nets be Universal Approximators for Fuzzy Functions?, Fuzzy Sets and Systems, 101(1999)323-330.

6. J.J. Buckley and Y. Qu: On Using α–cuts to Evaluate Fuzzy Equations, Fuzzy Sets and Systems, 38(1990)309-312.

7. P.T. Chang and E.S. Lee: Fuzzy Arithmetic and Comparison of Fuzzy Numbers, in: M. Delgado, J,Kacprzyk, J.L. Verdegay and M.A. Vila (eds.), Fuzzy Optimization: Recent Advances, Physica-Verlag, Heidelberg, Germany, 1994, 69-81.

8. S.J. Chen and C.L. Hwang: Fuzzy Multiple Attribute Decision Making, Springer-Verlag, Heidelberg, Germany, 1992.

9. D. Dubois, E. Kerre, R. Mesiar and H. Prade: Fuzzy Interval Analysis, in: D. Dubois and H. Prade (eds.), Fundamentals of Fuzzy Sets, The Handbook of Fuzzy Sets, Kluwer Acad. Publ., 2000, 483-581.

10. G.J. Klir and B. Yuan: Fuzzy Sets and Fuzzy Logic: Theory and Applications, Prentice Hall, Upper Saddle River, N.J., 1995.

11. V. Kreinovich, L. Longpre and J.J. Buckley: Are There Easy-to-Check Necessary and Sufficient Conditions for Straightforward Interval Computations to be Exact?, Reliable Computing, 9(2003)349-358.

12. R.E. Moore: Methods and Applications of Interval Analysis, SIAM Studies in Applied Mathematics, Philadelphia, 1979.

13. A. Neumaier: Interval Methods for Systems of Equations, Cambridge University Press, Cambridge, U.K., 1990.

14. X. Wang and E.E. Kerre: Reasonable Properties for the Ordering of Fuzzy Quantities (I), Fuzzy Sets and Systems, 118(2001)375-385.

15. X. Wang and E.E. Kerre: Reasonable Properties for the Ordering of Fuzzy Quantities (II), Fuzzy Sets and Systems, 118(2001)387-405.

Chapter 3

Fuzzy Probability Theory

3.1 Introduction

The first thing to do is explain how we will get fuzzy probabilities, which will be fuzzy numbers, from a set of confidence intervals. This is done in the next section. Then we discuss "restricted fuzzy arithmetic" in Section 3.4. Throughout this book whenever we wish to find the α-cut of a fuzzy probability, or a certain fuzzy number, we usually need to solve an optimization problem. We discuss this computation problem in more detail in Section 3.4.3. Then we turn to the development of the elementary properties of discrete (finite) fuzzy probability distributions in Sections 3.5-3.8. Some applications are in the last section Section 3.9. Fuzzy probability density/mass functions are in the next two chapters.

3.2 Fuzzy Probabilities from Confidence Intervals

We have an experiment in mind in which we are interested in only two possible outcomes labeled "success" and "failure". Let p be the probability of a success so that $q = 1 - p$ will be the probability of a failure. We want to estimate the value of p. We therefore gather a random sample which here is running the experiment n independent times and counting the number of times we had a success. Let x be the number of times we observed a success in n independent repetitions of this experiment. Then our point estimate of p is $\widehat{p} = x/n$.

We know that (Section 7.5 in [5]) that $(\widehat{p} - p)/\sqrt{p(1 - p)/n}$ is approximately $N(0, 1)$ if n is sufficiently large. Throughout this book we will always assume that the sample size is large enough for the normal approximation to

James J. Buckley: *Fuzzy Probability and Statistics*, StudFuzz **196**, 21 – 49 (2006)
www.springerlink.com © Springer-Verlag Berlin Heidelberg 2006

the binomial. Then

$$P(-z_{\beta/2} \leq \frac{\widehat{p} - p}{\sqrt{p(1-p)/n}} \leq z_{\beta/2}) \approx 1 - \beta, \qquad (3.1)$$

where $z_{\beta/2}$ was defined by

$$P(X \geq z_{\beta/2}) = \beta/2, \qquad (3.2)$$

for a $N(0,1)$ random variable X. Solving the inequality for the p in the numerator we have

$$P(\widehat{p} - z_{\beta/2}\sqrt{p(1-p)/n} \leq p \leq \widehat{p} + z_{\beta/2}\sqrt{p(1-p)/n}) \approx 1 - \beta. \qquad (3.3)$$

This leads to the $(1 - \beta)100\%$ approximate confidence interval for p

$$[\widehat{p} - z_{\beta/2}\sqrt{p(1-p)/n}, \widehat{p} + z_{\beta/2}\sqrt{p(1-p)/n}]. \qquad (3.4)$$

However, we have no value for p to use in this confidence interval. So, still assuming that n is sufficiently large, the usual procedure is to substitute $\widehat{p}$ for p in equation (3.4), using $\widehat{q} = 1 - \widehat{p}$, and we get the final $(1 - \beta)100\%$ approximate confidence interval

$$[\widehat{p} - z_{\beta/2}\sqrt{\widehat{pq}/n}, \widehat{p} + z_{\beta/2}\sqrt{\widehat{pq}/n}]. \qquad (3.5)$$

Assume that $0.01 \leq \beta < 1$. Staring at 0.01 is arbitrary and you could begin at 0.001 or 0.10 etc. In this book we usually start at 0.01. Denote these confidence intervals as

$$[p_1(\beta), p_2(\beta)], \qquad (3.6)$$

for $0.01 \leq \beta < 1$. Add to this the interval $[\widehat{p}, \widehat{p}]$ for the 0% confidence interval for p. Then we have a $(1 - \beta)100\%$ confidence interval for p for $0.01 \leq \beta \leq 1$.

Now place these confidence intervals, one on top of another, to produce a triangular shaped fuzzy number $\overline{p}$ whose alpha-cuts are the confidence intervals. We have

$$\overline{p}[\alpha] = [p_1(\alpha), p_2(\alpha)], \qquad (3.7)$$

for $0.01 \leq \alpha \leq 1$. All that is needed is to finish the "bottom" of $\overline{p}$ to make it a complete fuzzy number. We will simply drop the graph of $\overline{p}$ straight down to complete its α-cuts so

$$\overline{p}[\alpha] = [p_1(0.01), p_2(0.01)], \qquad (3.8)$$

for $0 \leq \alpha < 0.01$. In this way we are using more information in $\overline{p}$ than just a point estimate or just a single confidence interval. Then $\overline{p}$, a triangular shaped fuzzy number, will be our fuzzy estimator for p

The following example shows that $\overline{p}$ will be a triangular shaped fuzzy number. However, for simplicity, throughout this book we will always use triangular fuzzy numbers for the fuzzy values of uncertain parameters (like means, variances,etc.) in probability density (mass) functions.

The method used above to construct $\overline{p}$ will be used throughout the book in obtaining fuzzy estimators from data.

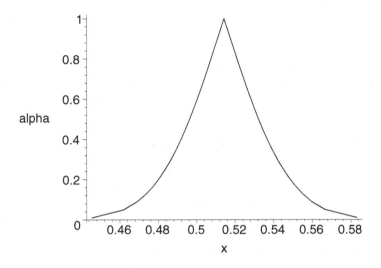

Figure 3.1: Fuzzy Estimator $\overline{p}$ in Example 3.2.1, $0.01 \leq \beta \leq 1$

Example 3.2.1

Assume that $n = 350$, $x = 180$ so that $\widehat{p} = 0.5143$. The confidence intervals become

$$[0.5143 - 0.0267 z_{\beta/2}, 0.5143 + 0.0267 z_{\beta/2}], \tag{3.9}$$

for $0.01 \leq \beta \leq 1$.

To obtain a graph of fuzzy p, or $\overline{p}$, we use $0.01 \leq \beta \leq 1$. We evaluated equation (3.9) using Maple [12] and then the graph of $\overline{p}$ is shown in Figure 3.1, without dropping the graph straight down to the x-axis at the end points. The Maple commands for this figure are similar to those for Figure 6.2 which are in Chapter 30.

To complete the picture we draw short vertical line segments, from the horizontal axis up to the graph, at the end points of the base of the fuzzy number $\overline{p}$. The base ($\overline{p}[0]$) in Figure 3.1 is a 99% confidence interval for p.

3.3   Fuzzy Probabilities from Expert Opinion

Some of the fuzzy probabilities and fuzzy constants in our models may have to be estimated by experts. That is, we have no statistical data to generate the fuzzy numbers from a set of confidence intervals as discussed in the previous section. So let us briefly see how this may be accomplished. First assume we have only one expert and he/she is to estimate the value of some probability p. We can solicit this estimate from the expert as is done in estimating job times in project scheduling ([18], Chapter 13). Let $a =$ the "pessimistic" value of p, or the smallest possible value, let $c =$ be the "optimistic" value of

p, or the highest possible value, and let $b =$ the most likely value of p. We then ask the expert to give values for a, b, c and we construct the triangular fuzzy number $\overline{p} = (a/b/c)$ for p. If we have a group of N experts all to estimate the value of p we solicit the a_i, b_i and c_i, $1 \leq i \leq N$, from them. Let a be the average of the a_i, b is the mean of the b_i and c is the average of the c_i. The simplest thing to do is to use $(a/b/c)$ for p.

3.4 Restricted Fuzzy Arithmetic

Restricted fuzzy arithmetic was first proposed in the papers ([6]-[9],[13]-[15], see also [11]). In these papers restricted fuzzy arithmetic due to probabilistic constraints is mentioned but was not developed to the extent that it will be in this book. Also, in [15] the authors extend the results in [14] to fuzzy numbers for probabilities under restricted fuzzy arithmetic due to probabilistic constraints similar to what we use in this book. But in [15] they concentrate only on Bayes' formula for updating prior fuzzy probabilities to posterior fuzzy probabilities.

In this section we will first discuss three methods that you may obtain crisp/fuzzy probabilities. Next we explain what we mean by restricted fuzzy arithmetic for a finite, discrete, fuzzy probability distribution ([1],[2]). Then we will extend it to how it will be used in this book. Restricted fuzzy arithmetic is used only in Chapters 3 and 4.

3.4.1 Probabilities

Let $X = \{x_1, ..., x_n\}$ be a finite set and let P be a probability function defined on all subsets of X with $P(\{x_i\}) = a_i$, $1 \leq i \leq n$, $0 < a_i < 1$, all i, and $\sum_{i=1}^{n} a_i = 1$. Starting in this chapter we will substitute a fuzzy number $\overline{a}_i$ for a_i, for some i, to obtain a discrete (finite) fuzzy probability distribution. Where do these fuzzy numbers come from?

In some problems, because of the way the problem is stated, the values of all the a_i are crisp and known. For example, consider tossing a fair coin and $a_1 =$ the probability of getting a "head" and $a_2 =$ is the probability of obtaining a "tail". Since we assumed it to be a fair coin we must have $a_1 = a_2 = 0.5$. In this case we would not substitute a fuzzy number for a_1 or a_2 because they are crisp numbers 0.5. But in many other problems the a_i are not known exactly and they are either estimated from a random sample or they are obtained from "expert opinion" .

Suppose we have the results of a random sample to estimate the value of a_1. We would construct a set of confidence intervals for a_1 and then put these together to get the fuzzy number $\overline{a}_1$ for a_1. This method of building a fuzzy number from confidence intervals was discussed in Section 3.2.

Assume that we do not know the values of the a_i and we do not have any data to estimate their values. Then we may obtain numbers for the a_i from some group of experts. This group could consist of only one expert. This case includes subjective, or "personal", probabilities and it is used to estimate certain probabilities and certain parameters in this book. We discussed this case in Section 3.3 above.

So, when we have to estimate probabilities/parameters from data/expert opinion, we will use fuzzy numbers for these uncertain items.

3.4.2 Restricted Arithmetic: General

Let $X = \{x_1, ..., x_n\}$ be a finite set and let P be a probability function defined on all subsets of X with $P(\{x_i\}) = a_i$, $1 \leq i \leq n$, $0 < a_i < 1$ all i and $\sum_{i=1}^{n} a_i = 1$. X together with P is a discrete (finite) probability distribution. In practice all the a_i values must be known exactly. Many times these values are estimated, or they are provided by experts. We now assume that some of these a_i values are uncertain and we will model this uncertainty using fuzzy numbers. Not all the a_i need to be uncertain, some may be known exactly and are given as a crisp (real) number. If an a_i is crisp, then we will still write it as a fuzzy number even though this fuzzy number is crisp.

Due to the uncertainty in the a_i values we substitute $\overline{a}_i$, a fuzzy number, for each a_i and assume that $0 < \overline{a}_i < 1$ all i. If some a_i is known precisely, then this $\overline{a}_i = a_i$ but we still write a_i as $\overline{a}_i$. Then X together with the $\overline{a}_i$ values is a discrete (finite) fuzzy probability distribution. We write $\overline{P}$ for fuzzy P and we have $\overline{P}(\{x_i\}) = \overline{a}_i$, $1 \leq i \leq n$, $0 < \overline{a}_i < 1$.

The uncertainty is in some of the a_i values but we know that we have a discrete probability distribution. So we now put the following restriction on the $\overline{a}_i$ values: there are $a_i \in \overline{a}_i[1]$ so that $\sum_{i=1}^{n} a_i = 1$. That is, we can choose a_i in $\overline{a}_i[\alpha]$, all α, so that we get a discrete probability distribution.

Let A and B be (crisp) subsets of X. We know how to compute $P(A)$ and $P(B)$ so let us find $\overline{P}(A)$ and $\overline{P}(B)$, the fuzzy probabilities of A and B, respectively. To do this we introduce restricted fuzzy arithmetic. There may be uncertainty in some of the a_i values, but there is no uncertainty in the fact that we have a discrete probability distribution. That is, whatever the a_i values in $\overline{a}_i[\alpha]$ we must have $a_1 + ... + a_n = 1$. This is the basis of our restricted fuzzy arithmetic. Suppose $A = \{x_1, ..., x_k\}$, $1 \leq k < n$, then define (this is an α-cut of the fuzzy probability)

$$\overline{P}(A)[\alpha] = \{\sum_{i=1}^{k} a_i | \ \mathbf{S} \ \}, \tag{3.10}$$

for $0 \leq \alpha \leq 1$, where $\mathbf{S}$ stands for the statement "$a_i \in \overline{a}_i[\alpha]$, $1 \leq i \leq n$, $\sum_{i=1}^{n} a_i = 1$ ". This is our restricted fuzzy arithmetic. Notice that we first choose a complete discrete probability distribution from the α-cuts before we compute a probability in equation (3.10). Notice also that $\overline{P}(A)[\alpha]$ is not the sum of the intervals $\overline{a}_i[\alpha]$, $1 \leq i \leq k$, using interval arithmetic because this sum can many times exceed one. The α-cuts defined in equation (3.10) are then put together to obtain a fuzzy number for the fuzzy probability. $\overline{P}(A)[\alpha]$ will be an interval, and we will use the right side of equation (3.10) to compute the end points of this interval as in equations (2.27) and (2.28) in Chapter 2.

3.4.3 Computing Fuzzy Probabilities

Throughout this book whenever we wish to find the α-cut of a fuzzy probability we will need to solve an optimization problem. The problem is to find the max and min of a function $f(p_1, ..., p_n)$ subject to linear constraints. There will be two types of problems. The first one, needed in this chapter and Chapter 4 will be described in the next subsection (called the "First Problem") and the other type of problem, used in Chapters 4 and 5, is discussed in the second subsection (called the "Second Problem").

First Problem

The structure of this problem is

$$max/min f(p_{i_1}, ..., p_{i_K}) \qquad (3.11)$$

subject to

$$a_i \leq p_i \leq b_i, 1 \leq i \leq n, \qquad (3.12)$$

and

$$p_1 + ... + p_n = 1. \qquad (3.13)$$

The set $\{p_{i_1}, ..., p_{i_K}\}$ is a subset of $\{p_1, ..., p_n\}$. The p_i must be in interval $[a_i, b_i]$, $1 \leq i \leq n$ and their sum must be one. In the application of this problem : (1) the p_i will be probabilities ; (2) the intervals $[a_i, b_i]$ will be α-cuts of fuzzy numbers used for fuzzy probabilities; and (3) the sum of the p_i equals one means the sum of the probabilities is one. This problem is actually solving for the α-cuts of a fuzzy probability.

In this subsection we explain how we will obtain numerical solutions to this problem. If $f(p_{i_1}, ..., p_{i_K})$ is a linear function of the $p_i's$, then this problem is a linear programming problem and we can solve it using the "Optimization" call in Maple [12]. So now assume that the function f is not a linear function of the $p_i's$. We used three methods of solution: (1)graphical; (2) calculus; and (3) Maple/Solver. The calculus procedure is discussed in the three examples below, so now let us first discuss the graphical method. Solver ([4],[17]) is nonlinear optimization software which is a free add on to Microsoft Excel. Solver was used mostly in Chapter 13. Maple can solve many nonlinear optimization problems with the "NLPSolve" command.

The graphical method is applicable for $n = 2$, and sometimes for $n = 3$. First let $n = 2$ and assume that $\overline{p}_2 = 1 - \overline{p}_1$ so that we may substitute $1 - p_1$ for p_2 in the function f and obtain f a function of p_1 only. The optimization problem is now $max/min f(p_1)$ subject to $a_1 \leq p_1 \leq b_1$. We simply used Maple to graph $f(p_1)$ for $a_1 \leq p_1 \leq b_1$. From the graph we can sometimes easily find the max, or min, especially if they are at the end points. Now suppose the max (min) is not at an end point. Then we repeatedly evaluated the function in the neighborhood of the extreme point until we could estimate its value to our desired accuracy. Next let $n = 3$ and assume that $\bar{p}_3 = 1 - (\bar{p}_1 + \bar{p}_2)$. For example $\bar{p}_1 = (0.2/0.3/0.4) = \bar{p}_2$ and $\bar{p}_3 = (0.2/0.4/0.6)$ satisfies this constraint. Then substitute $1 - (p_1 + p_2)$ for p_3 in f to get f a function of only p_1 and p_2. Then the optimization problem becomes $max/min f(p_1, p_2)$ subject to $a_1 \leq p_1 \leq b_1$, $a_2 \leq p_2 \leq b_2$. We used Maple to graph the surface over the rectangle $a_i \leq p_i \leq b_i$, $i = 1, 2$. Then, as in the $n = 2$ case, we found by inspection extreme points, or we used Maple to repeatedly evaluate the function in the neighborhood of an extreme point to estimate the max, or min, value. This method of looking at the surface $z = f(p_1, p_2)$ is also applicable when $n = 2$ and we can not use $\bar{p}_2 = 1 - \bar{p}_1$.

Now let us look at three examples of the calculus method. But first we need to define what we will mean by saying that a certain subset of the p_i, $1 \leq i \leq n$, is feasible. To keep the notation simple let $n = 5$ and we claim that p_1, p_2 and p_4 are feasible. This means that we may choose any $p_i \in \bar{p}_i[\alpha]$, $i = 1, 2, 4$, and then we can then find a $p_3 \in \bar{p}_3[\alpha]$ and a $p_5 \in \bar{p}_5[\alpha]$ so that $p_1 + p_2 + p_3 + p_4 + p_5 = 1$. Assume that f is a function of only p_1, p_2 and p_4. Also assume that f is: (1) an increasing function of p_1 and p_4; and (2) a decreasing function of p_2. We will be using $[a_i, b_i] = \bar{p}_i[\alpha]$, $1 \leq i \leq 5$. If p_1, p_2 and p_4 are feasible, then, as in the examples below, we may find that for $\alpha \in [0, 1]$: (1) $min f(p_1, p_2, p_4) = f(a_1, b_2, a_4)$; and (2) $max f(p_1, p_2, p_4) = f(b_1, a_2, b_4)$.

Example 3.4.3.1

Consider the problem

$$max/min f(p_1, p_2) \tag{3.14}$$

subject to

$$p_1 \in [a_1, b_1], p_2 \in [a_2, b_2], p_1 + p_2 = 1, \tag{3.15}$$

where $0 \leq a_i \leq b_i \leq 1$, $i = 1, 2$. Also assume that $\partial f/\partial p_1 > 0$ and $\partial f/\partial p_2 < 0$ for the p_i in $[0, 1]$. Now the result depends on the two intervals $[a_1, b_1]$ and $[a_2, b_2]$.

First assume that $1 - a_1 = b_2$ and $1 - b_1 = a_2$. This means that $p_1 = a_1$ and $p_2 = b_2$ are feasible because $a_1 + b_2 = 1$. Also, $p_1 = b_1$ and $p_2 = a_2$ are feasible since $b_1 + a_2 = 1$. For example, $[0.3, 0.6]$ and $[0.4, 0.7]$ are two such intervals. Then

$$min f(p_1, p_2) = f(a_1, b_2), \tag{3.16}$$

and

$$max f(p_1, p_2) = f(b_1, a_2). \tag{3.17}$$

Now assume that $1 - a_1$ does not equal b_2 or $1 - b_1$ is not equal to a_2. Then the optimization problem is not so simple and we may try to use the graphical procedure presented above, graphing the surface $f(p_1, p_2)$ over the rectangle $a_1 \leq p_i \leq b_i$ for $i = 1, 2$, to approximate $max/minf(p_1, p_2)$. We might also employ Maple or Solver.

Example 3.4.3.2

Now we have the problem

$$max/minf(p_1, p_2, p_3),   \tag{3.18}$$

subject to

$$p_i \in [a_i, b_i], 1 \leq i \leq 3, p_1 + p_2 + p_3 = 1,   \tag{3.19}$$

where $0 \leq a_i \leq b_i \leq 1$, $i = 1, 2, 3$. Also assume that $\partial f/\partial p_1 > 0$, $\partial f/\partial p_2 < 0$ and $\partial f/\partial p_3 < 0$. If $a_1 + b_2 + b_3 = 1$ and $b_1 + a_2 + a_3 = 1$, then the solution is

$$minf(p_1, p_2, p_3) = f(a_1, b_2, b_3),   \tag{3.20}$$

and

$$maxf(p_1, p_2, p_3) = f(b_1, a_2, a_3).   \tag{3.21}$$

When these sums do not add up to one, we need to employ some numerical optimization method like Maple or Solver.

Example 3.4.3.3

The last problem is

$$max/minf(p_1, p_3),   \tag{3.22}$$

subject to

$$p_i \in [a_i, b_i], 1 \leq i \leq 3, p_1 + p_2 + p_3 = 1,   \tag{3.23}$$

where $0 \leq a_i \leq b_i \leq 1$, $1 \leq i \leq 3$. Assume that $\partial f/\partial p_1 > 0$ and $\partial f/\partial p_3 < 0$. Also assume that: (1)$p_1 = a_1$ and $p_3 = b_3$ are feasible, or $a_1 + p_2 + b_3 = 1$ for some $p_2 \in [a_2, b_2]$; and (2) $p_1 = b_1$ and $p_3 = a_3$ are feasible, or $b_1 + p_2 + a_3 = 1$ for some $p_2 \in [a_2, b_2]$. Then the solution is

$$minf(p_1, p_3) = f(a_1, b_3),   \tag{3.24}$$

and

$$maxf(p_1, p_3) = f(b_1, a_3).   \tag{3.25}$$

We first try the calculus method and if that procedure is not going to work, then we next try the graphical method. The graphical, or calculus, method was applicable in many types of problems in [1]. However, one can easily consider situations where neither procedure is applicable. Then we will use Maple or Solver.

New Algorithm

When the objective function is linear we have a new algorithm [16] that many be used. We do not guarantee that it will always work but it is easy to employ using just pencil and paper. This method may be used to solve various optimization problems, without the need of "Optimization" in Maple, in [1] but we shall present only the following two examples. The first simple example is from the first page in that book.

Example 3.4.3.4

Consider

$$max/min(a_1 + a_2),\qquad(3.26)$$

subject to

$$a_1 \in [0.1, 0.3], a_2 \in [0.3, 0.7], a_3 \in [0.2, 0.4], a_1 + a_2 + a_3 = 1.\qquad(3.27)$$

First look at the maximum. Give the a_i their minimum values $a_1 = 0.1$, $a_2 = 0.3$, $a_3 = 0.2$. Then $a_1 + a_2 + a_3 = 0.6$ and 0.4 must be added to the a_i. The remaining 0.4 is added to maximize $a_1 + a_2$. The order is a_1, a_2, a_3 or a_2, a_1, a_3. First consider a_1, a_2, a_3. Now a_1 takes 0.2 to get to its max. of 0.3 and the remaining 0.2 is given to a_2 making its value 0.5. The max. is 0.8 for $a_1 = 0.3$, $a_2 = 0.5$ and $a_3 = 0.2$. With the order a_2, a_1, a_3 we first give all of 0.4 to a_2 making it 0.7 and $a_1 = 0.1$, $a_2 = 0.7$, $a_3 = 0.2$ for an alternate max. of 0.8.

Next consider the minimum. Give the minimum values to the a_i and the remaining 0.4 is given to the a_i in the order a_3, a_1, a_2 or a_3, a_2, a_1. Assume the first ordering. Give the max. 0.2 to a_3 to make its value 0.4 and the remaining 0.2 goes to a_2 making $a_2 = 0.5$. The minimum is 0.6 for $a_1 = 0.1$, $a_2 = 0.5$ and $a_3 = 0.4$. An alternate minimum solution may exist using the other ordering. Maple commands to solve this problem are given in Chapter 30.

Example 3.4.3.5

This problem is from the "Blood Types" example in Section 3.9.1 in this chapter. We want to maximize $p_a + p_b + p_{ab}$ subject to $0.3 + 0.03\alpha \le p_a \le 0.36 - 0.03\alpha$, $0.2 + 0.03\alpha \le p_b \le 0.26 - 0.03\alpha$, $0.32 + 0.03\alpha \le p_{ab} \le 0.38 - 0.03\alpha$, $0.06 + 0.03\alpha \le p_0 \le 0.12 - 0.03\alpha$, $p_a + p_b + p_{ab} + p_0 = 1$. First set $p_a = 0.3 + 0.03\alpha$, $p_b = 0.2 + 0.03\alpha$, $p_{ab} = 0.32 + 0.03\alpha$, $p_0 = 0.06 + 0.03\alpha$. Then $p_a + p_b + p_{ab} + p_0 = 0.88 + 0.12\alpha$. The amount to be added to get sum one is $0.12 - 0.12\alpha$. The order will be p_a, p_b, p_{ab} and p_0. p_a is raised with $0.06 - 0.06\alpha$ to $0.36 - 0.03\alpha$. Remaining to be added is $0.06 - 0.06\alpha$. This amount is added to p_b which now equals $0.26 - 0.03\alpha$. Then the maximum of

$p_a + p_b + p_{ab}$ is $(0.36 - 0.03\alpha) + (0.26 - 0.03\alpha) + (0.32 + 0.03\alpha) = 0.94 - 0.03\alpha$. In a similar manner, by raising p_0 to its maximum first as in Example 3.4.3.4, we find the minimum to be $0.88 + 0.03\alpha$. We leave this part as an exercise for the interested reader.

Second Problem

This type of problem is

$$max/min f(\theta),   \tag{3.28}$$

subject to

$$\theta_i \in [a_i, b_i], 1 \le i \le n,   \tag{3.29}$$

where $\theta = (\theta_1, ..., \theta_n)$. Notice in this case we do not have the constraint that the θ_i must sum to one. In applications of this problem the intervals $[a_i, b_i]$ are α-cuts of fuzzy numbers used for uncertain parameters in probability density (mass) functions. This problem is actually solving to obtain the α-cuts of a fuzzy probability.

In this subsection we explain how we will obtain numerical solutions to this problem. In this book n will be one or two, and n will be two only for the normal probability density function. When $n = 1$ we may employ a calculus method (Example 3.4.3.6 below) or a graphical procedure (discussed in the previous subsection). When $n = 2$ we used the graphical method (see Example 5.3.1). To see more detail on this type of problem let us look at the next three examples. Because of the nature of the objective function in the first two examples our nonlinear solution software Solver was not applicable. However, Maple can be used.

Example 3.4.3.6

Let $N(\mu, \sigma^2)$ be the normal probability density with mean μ and variance σ^2. To obtain the fuzzy normal we use fuzzy numbers $\overline{\mu}$ and $\overline{\sigma}^2$ for μ and σ^2, respectively. Set $\overline{P}[c, d]$ to be the fuzzy probability of obtaining a value in the interval $[c, d]$. Its α-cuts are gotten by solving the following problem (see Section 5.3)

$$max/min f(\mu, \sigma^2) = \int_{z_1}^{z_2} N(0, 1) dx,   \tag{3.30}$$

subject to

$$\mu \in [a_1, b_1], \sigma^2 \in [a_2, b_2],   \tag{3.31}$$

where $a_i \le b_i$, $1 \le i \le 2$, $a_2 > 0$, and $z_1 = (d - \mu)/\sigma$ and $z_2 = (c - \mu)/\sigma$, and $N(0, 1)$ is the normal with zero mean and unit variance. We use the graphical method, discussed above, to solve this problem in Example 5.3.1. We may also use "NLPSolve" in Maple to solve this problem (see Example 5.3.1 and the Maple commands for Example 5.3.1 are in Chapter 30).

Example 3.4.3.7

The negative exponential has density $f(x; \lambda) = \lambda \exp(-\lambda x)$ for $x \geq 0$, and the density is zero for $x < 0$. The fuzzy negative exponential has a fuzzy number, say $\overline{\lambda} = (2/4/6)$, substituted for crisp λ. We wish to calculate the fuzzy probability of obtaining a value in the interval $[6, 10]$. Let this fuzzy probability be $\overline{P}[6, 10]$ and its α-cuts , see Section 5.4, are determined from the following problem

$$max/min f(\lambda) = \int_6^{10} \lambda \exp(-\lambda x) dx, \tag{3.32}$$

subject to

$$\lambda \in [a, b], \tag{3.33}$$

where $[a, b]$ will be an α-cut of $(2/4/6)$. This problem is easy to solve because $f(\lambda)$ is a decreasing function of λ, $df/d\lambda < 0$, across the interval $[a, b]$ (which is a subset of $[2, 6]$). Hence,

$$min f(\lambda) = f(b), \tag{3.34}$$

and

$$max f(\lambda) = f(a). \tag{3.35}$$

Maple commands for this problem are in Chapter 30.

Example 3.4.3.8

Here we have a fuzzy, discrete, probability distribution $\overline{p}_i$, $0 \leq i \leq M$, over $X = \{0, 1, 2, ..., M\}$. We want to find the mean of this fuzzy probability distribution. The mean $\overline{\mu}$ is

$$\overline{\mu} = \sum_{k=0}^{M} k \overline{p}_k, \tag{3.36}$$

which is evaluated by α-cuts and restricted fuzzy arithmetic. So

$$\overline{\mu}[\alpha] = \{\sum_{k=0}^{M} k p_k | \ \mathbf{S} \ \}, \tag{3.37}$$

where $\mathbf{S}$ is "$p_i \in \overline{p}_i[\alpha]$, $0 \leq i \leq M$, and $p_0 + ... + p_M = 1$". Let $\overline{\mu}[\alpha] = [\mu_1(\alpha), \mu_2(\alpha)]$. The optimization problems are

$$\mu_1(\alpha) = min \sum_{k=0}^{M} k p_k, \tag{3.38}$$

subject to the constraints in statement **S**, and

$$\mu_2(\alpha) = max \sum_{k=0}^{M} kp_k, \tag{3.39}$$

subject to the same constraints.

Both optimization problems are linear programming problems and hence can be solved using Maple [12].

3.5   Fuzzy Probability

Let A and B be (crisp) subsets of $X = \{x_1, ..., x_n\}$. We have a discrete (finite) fuzzy probability distribution $\overline{P}(\{x_i\}) = \overline{a}_1$, $0 < \overline{a}_i < 1$, $1 \leq i \leq n$. Suppose $A = \{x_1, ..., x_k\}$, $1 \leq k < n$, then define

$$\overline{P}(A)[\alpha] = \{\sum_{i=1}^{k} a_i| \quad \mathbf{S} \quad \}, \tag{3.40}$$

for $0 \leq \alpha \leq 1$, where **S** stands for the statement "$a_i \in \overline{a}_i[\alpha]$, $1 \leq i \leq n$, $\sum_{i=1}^{n} a_i = 1$". This is restricted fuzzy arithmetic. We now show that the $\overline{P}(A)[\alpha]$ are the α-cuts of a fuzzy number $\overline{P}(A)$. But first we require some definitions.

Define

$$S = \{(x_1, ..., x_n)|x_i \geq 0 \; all \; i, \sum_{i=1}^{n} x_i = 1\}, \tag{3.41}$$

and then also define

$$Dom[\alpha] = (\prod_{i=1}^{n} \overline{a}_i[\alpha]) \bigcap S, \tag{3.42}$$

for $0 \leq \alpha \leq 1$. In equation (3.42) we first take the product of n closed intervals producing a "rectangle" in n dimensional space which is then intersected with the set S. Now define a function f mapping $Dom[\alpha]$ into the real numbers as

$$f(a_1, ..., a_n) = \sum_{i=1}^{k} a_i, \tag{3.43}$$

for $(a_1, ..., a_n) \in Dom[\alpha]$. f is continuous, $Dom[\alpha]$ is connected, closed, and bounded implying that the range of f is a closed, and bounded, interval of real numbers. Define

$$\Gamma[\alpha] = f(Dom[\alpha]), \tag{3.44}$$

for $0 \leq \alpha \leq 1$. But, from equation (3.40), we see that

$$\overline{P}(A)[\alpha] = \Gamma[\alpha], \tag{3.45}$$

for all α. Hence, $\overline{P}(A)$ is a fuzzy number since it is normalized ($\overline{P}(A)[1] \neq \phi$). We are using the general definition of fuzzy numbers ([3],[10]).

We can now argue that:

1. If $A \cap B = \phi$, then $\overline{P}(A) + \overline{P}(B) \geq \overline{P}(A \cup B)$.

2. If $A \subseteq B$ and $\overline{P}(A)[\alpha] = [p_{a1}(\alpha), p_{a2}(\alpha)]$ and $\overline{P}(B)[\alpha] = [p_{b1}(\alpha), p_{b2}(\alpha)]$, then $p_{ai}(\alpha) \leq p_{bi}(\alpha)$ for $i = 1, 2$ and $0 \leq \alpha \leq 1$.

3. $0 \leq \overline{P}(A) \leq 1$ all A with $\overline{P}(\phi) = 0$, $\overline{P}(X) = 1$.

4. $\overline{P}(A) + \overline{P}(A') \geq 1$, where A' is the complement of A.

5. When $A \cap B \neq \phi$, $\overline{P}(A \cup B) \leq \overline{P}(A) + \overline{P}(B) - \overline{P}(A \cap B)$.

It is easy to see that (2) and (3) are true and (4) follows from (1) and (3). So we now demonstrate that (1) and the generalized addition law (5) are true. Then we show by Example 3.5.1 below that in cases (1) and (5) we may not get equality.

We show that $(\overline{P}(A) + \overline{P}(B))[\alpha] = \overline{P}(A)[\alpha] + \overline{P}(B)[\alpha] \supseteq \overline{P}(A \cup B)[\alpha]$, for all α. To simplify the discussion assume that $A = \{x_1, ..., x_k\}$, $B = \{x_l, ..., x_m\}$ for $1 \leq k < l \leq m \leq n$. Again let **S** denote the statement "$a_i \in \overline{a}_i[\alpha]$, $1 \leq i \leq n$, $\sum_{i=1}^{n} a_i = 1$". Then we need to show, based on equation (3.10), that

$$\{\sum_{i=1}^{k} a_i | \; \mathbf{S} \;\} + \{\sum_{i=l}^{m} a_i | \; \mathbf{S} \;\} \supseteq \{\sum_{i=1}^{k} a_i + \sum_{i=l}^{m} a_i | \; \mathbf{S} \;\}. \qquad (3.46)$$

Let $r = s + t$ be a member of the right side of equation (3.46) where $s = a_1 + ... + a_k$ and $t = a_l + ... + a_m$. Then s belongs to the first member of the left side of equation (3.46) and t belongs to the second member. Hence $r = s + t$ belongs to the left side of equation (3.46) and equation (3.46) is correct.

However, there are situations where $\overline{P}(A) + \overline{P}(B) = \overline{P}(A \cup B)$ when A and B are disjoint. We also give an example of equality in Example 3.5.1 below.

Next we wish to show (5) is also true. Using the notation defined above assume that $A = \{x_1, ..., x_k\}$, $B = \{x_l, ..., x_m\}$ but now $1 \leq l \leq k \leq m \leq n$. We show that $\overline{P}(A)[\alpha] + \overline{P}(B)[\alpha] - \overline{P}(A \cap B)[\alpha] \supseteq \overline{P}(A \cup B)[\alpha]$. Or, we show that

$$\{\sum_{i=1}^{k} a_i | \mathbf{S}\} + \{\sum_{i=l}^{m} a_i | \mathbf{S}\} - \{\sum_{i=l}^{k} a_i | \mathbf{S}\} \supseteq \{\sum_{i=1}^{m} a_i | \mathbf{S}\}. \qquad (3.47)$$

Let r be in the right side of equation (3.47). Then we may write r as $r = s + t - u$ where $s = a_1 + ... + a_k$, $t = a_l + ... + a_m$ and $u = a_l + ... + a_k$. Now s belongs to the first member on the left side of equation (3.47), t belongs to the second member and u belongs to the third member. Hence r belongs to the left side of equation (3.47).

Example 3.5.1

We first show by example that you may not obtain equality in equation (3.46). Let $n = 5$, $A = \{x_1, x_2\}$, $B = \{x_4, x_5\}$, $a_i = 0.2$ for $1 \leq i \leq 5$. All the probabilities are uncertain except a_3 so let $\overline{a}_1 = \overline{a}_2 = (0.19/0.2/0.21)$, $\overline{a}_3 = 0.2$ and $\overline{a}_4 = \overline{a}_5 = (0.19/0.2/0.21)$. Then $\overline{P}(A)[0] = [0.38, 0.42]$ because $p_1 = 0.19 = p_2$ are feasible (see Section 3.4.3)and $p_1 = p_2 = 0.21$ are feasible. We also determine $\overline{P}(B)[0] = [0.38, 0.42]$ so the left side of equation (3.46), for $\alpha = 0$, is the interval $[0.76, 0.84]$. However, $\overline{P}(A \cup B)[0] = [0.8, 0.8]$. For A and B disjoint, we can get $\overline{P}(A \cup B)[\alpha]$ a proper subset of $\overline{P}(A)[\alpha] + \overline{P}(B)[\alpha]$.

Let $n = 6$, $A = \{x_1, x_2, x_3\}$, $B = \{x_3, x_4, x_5\}$, $a_i = 0.1$ for $1 \leq i \leq 5$ and $a_6 = 0.5$, Assuming all probabilities are uncertain we substitute $\overline{a}_i = (0.05/0.1/0.15)$ for $1 \leq i \leq 5$ and $\overline{a}_6 = (0.25/0.5/0.75)$. Then we easily deduce that $\overline{P}(A \cup B)[0] = [0.25, 0.75]$, $\overline{P}(A)[0] = \overline{P}(B)[0] = [0.15, 0.45]$ and $\overline{P}(A \cap B)[0] = [0.05, 0.15]$. Then, from interval arithmetic, we see that

$$[0.25, 0.75] \neq [0.15, 0.45] + [0.15, 0.45] - [0.05, 0.15], \qquad (3.48)$$

where the right side of this equation is the interval $[0.15, 0.85]$. So $\overline{P}(A \cup B)[\alpha]$ can be a proper subset of $\overline{P}(A)[\alpha] + \overline{P}(B)[\alpha] - \overline{P}(A \cap B)[\alpha]$.

Now we show by example we can obtain $\overline{P}(A) + \overline{P}(B) = \overline{P}(A \cup B)$ when A and B are disjoint. Let $X = \{x_1, x_2, x_3\}$, $A = \{x_1\}$, $B = \{x_3\}$, $\overline{a}_1 = (0.3/0.33/0.36)$, $\overline{a}_2 = (0.28/0.34/0.40)$ and $\overline{a}_3 = \overline{a}_1$. Then $\overline{P}(A) = \overline{a}_1$, $\overline{P}(B) = \overline{a}_3$ so $\overline{P}(A) + \overline{P}(B) = (0.6/0.66/0.72)$. Alpha-cuts of $\overline{P}(A \cup B)$ are

$$\overline{P}(A \cup B)[\alpha] = \{a_1 + a_3 | \quad \mathbf{S} \quad \}. \qquad (3.49)$$

Let $\overline{a}_i[\alpha] = [a_{i1}(\alpha), a_{i2}(\alpha)]$, for $i = 1, 2, 3$. We can evaluate equation (3.49) by using the end points of these α-cuts because; (1) for any α there is a $a_2 \in \overline{a}_2[\alpha]$ so that $a_{11}(\alpha) + a_2 + a_{31}(\alpha) = 1$; and (2) for each α there is a $a_2 \in \overline{a}_2[\alpha]$ so that $a_{12}(\alpha) + a_2 + a_{32}(\alpha) = 1$. Then

$$\overline{P}(A \cup B)[\alpha] = [a_{11}(\alpha) + a_{31}(\alpha), a_{12}(\alpha) + a_{32}(\alpha)], \qquad (3.50)$$

so that $\overline{P}(A \cup B) = (0.6/0.66/0/72)$.

We will finish this section with the calculation of the mean and variance of a discrete fuzzy probability distribution. The fuzzy mean is defined by its α-cuts

$$\overline{\mu}[\alpha] = \{\sum_{i=1}^{n} x_i a_i | \quad \mathbf{S} \quad \}, \qquad (3.51)$$

where, as before, $\mathbf{S}$ denotes the statement " $a_i \in \overline{a}_i[\alpha], 1 \leq i \leq n, \sum_{i=1}^{n} a_i = 1$". The variance is also defined by its α-cuts as

$$\overline{\sigma}^2[\alpha] = \{\sum_{i=1}^{n} (x_i - \mu)^2 a_i | \quad \mathbf{S}, \mu = \sum_{i=1}^{n} x_i a_i\}. \qquad (3.52)$$

The mean $\bar{\mu}$ and variance $\bar{\sigma}^2$ will be fuzzy numbers because $\bar{\mu}[\alpha]$ and $\bar{\sigma}^2[\alpha]$ are closed, bounded, intervals for $0 \leq \alpha \leq 1$. The same proof, as was given for $\overline{P}(A)[\alpha]$, can be used to justify these statements.

Example 3.5.2

Let $X = \{0, 1, 2, 3, 4\}$ with $a_0 = a_4 = \frac{1}{16}$, $a_1 = a_3 = 0.25$ and $a_2 = \frac{3}{8}$. Then $\mu = 2$ and $\sigma^2 = 1$. Assume there is uncertainty only in a_1 and a_3 so we substitute $\bar{a}_1$ for a_1 and $\bar{a}_3$ for a_3. Let us use $\bar{a}_1 = \bar{a}_3 = (0.2/0.25/0.3)$. First compute $\bar{\mu}[\alpha]$. Use the numerical values for the x_i, a_0, a_2 and a_4 but choose $a_1 \in \bar{a}_1[\alpha]$ and $a_3 = 0.5 - a_1$ in $\bar{a}_3[\alpha]$ so that the sum of the a_i equals one. Then the formula for crisp $\mu = f_1(a_1) = 2.5 - 2a_1$ is a function of only a_1. We see that $\partial f_1/\partial a_1 < 0$. This allows us to compute the end points of the interval $\bar{\mu}[\alpha]$ which gives $[1.9 + 0.1\alpha, 2.1 - 0.1\alpha] = \bar{\mu}[\alpha]$, so that $\bar{\mu} = (1.9/2/2.1)$, a triangular fuzzy number. Since $\bar{a}_1[\alpha] = [0.2 + 0.05\alpha, 0.3 - 0.05\alpha]$, we used $0.3 - 0.05\alpha$ to get $1.9 + 0.1\alpha$ and $0.2 + 0.05\alpha$ to obtain $2.1 - 0.1\alpha$. We do the same with the crisp formula for σ^2 and we deduce that $\sigma^2 = f_2(a_1) = 0.75 + 2a_1 - 4a_1^2$, for a_1 in $\bar{a}_1[\alpha]$. If $\bar{\sigma}^2[\alpha] = [\sigma_1^2(\alpha), \sigma_2^2(\alpha)]$ we determine from $f_2(a_1)$ that $\sigma_1^2(\alpha) = f_2(0.2 + 0.05\alpha)$ but $\sigma_2^2(\alpha) = 1$ all α. So the α-cuts of the fuzzy variance are $[0.99 + 0.02\alpha - 0.01\alpha^2, 1]$, $0 \leq \alpha \leq 1$. The uncertainty in the variance is that it can be less than one but not more than one. The graph of the fuzzy variance is shown in Figure 3.2.

It can be computationally difficult, in general, to compute the intervals $\overline{P}(A)[\alpha]$ (equation (3.40)), $\bar{\mu}[\alpha]$ (equation (3.51)), and $\bar{\sigma}^2[\alpha]$ (equation (3.52)). All we need to do is to determine the end points of these intervals which can be written as a non-linear optimization problem. For example, for $\bar{\sigma}^2[\alpha]$, we have

$$\sigma_1^2(\alpha) = min\{\sum_{i=1}^{n}(x_i - \mu)^2 a_i| \quad \mathbf{S} \quad \},\qquad(3.53)$$

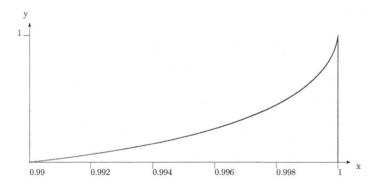

Figure 3.2: Fuzzy Variance in Example 3.5.2

and

$$\sigma_2^2(\alpha) = max\{\sum_{i=1}^{n}(x_i - \mu)^2 a_i| \quad \mathbf{S} \quad \},\tag{3.54}$$

where $\mathbf{S}$ is the statement "$a_i \in \overline{a}_i[\alpha], 1 \leq i \leq n, \mu = \sum_{i=1}^{n} x_i a_i, \sum_{i=1}^{n} a_i = 1$".
One may consider using Maple to estimate the $\sigma_i^2(\alpha)$ values for selected α.

3.6 Fuzzy Conditional Probability

Let $A = \{x_1, ..., x_k\}$, $B = \{x_l, ..., x_m\}$ for $1 \leq l \leq k \leq m \leq n$ so that A and
B are not disjoint. We wish to define the fuzzy conditional probability of
A given B. We will write this fuzzy conditional probability as $\overline{P}(A|B)$. We
now present two definitions for fuzzy conditional probability and then argue
in favor of the first definition.

Our first definition is

$$\overline{P}(A|B)[\alpha] = \{\frac{\sum_{i=l}^{k} a_i}{\sum_{i=l}^{m} a_i}| \quad \mathbf{S} \quad \}.\tag{3.55}$$

In this definition the numerator of the quotient is the sum of the a_i in the
intersection of A and B, while the denominator is the sum of the a_i in B.

Our second definition is

$$\overline{P}(A|B) = \frac{\overline{P}(A \cap B)}{\overline{P}(B)}.\tag{3.56}$$

This second definition seems very natural but, as to be shown in Example
3.6.1 below, because of fuzzy arithmetic this conditional fuzzy probability can
get outside the interval $[0, 1]$. The first definition always produces a fuzzy
probability in $[0, 1]$.

Example 3.6.1

Let $n = 4$, $A = \{x_1, x_2\}$, $B = \{x_2, x_3\}$ and all the a_i are uncertain with $\overline{a}_1 = (0.1/0.2/0.3)$, $\overline{a}_2 = (0.2/0.3/0.4)$, $\overline{a}_3 = (0/0.1/0.2)$ and $\overline{a}_4 = (0.3/0.4/0.5)$.
We show that, using the second definition, $\overline{P}(A|B)$ is not entirely in $[0, 1]$

Since $A \cap B = \{x_2\}$ we find that $\overline{P}(A \cap B) = \overline{a}_2$. We also easily find
that $\overline{P}(B) = (0.2/0.4/0.6)$. Then this definition produces a fuzzy number
$[(0.2/0.3/0.4)]/[(0.2/0.4/0.6)]$ whose $\alpha = 0$ cut is the interval $[\frac{1}{3}, 2]$ with right
end point greater than one. We would expect this to occur quite often using
the second definition so we drop the second definition and adopt the first
definition for fuzzy conditional probability.

Example 3.6.2

We will use the same data as given in Example 3.6.1 in the first definition to compute the fuzzy conditional probability of A given B. We need to evaluate

$$\overline{P}(A|B)[\alpha] = \{\frac{a_2}{a_2 + a_3}| \ \mathbf{S} \ \}, \tag{3.57}$$

for $0 \leq \alpha \leq 1$. If we let $y = f(a_2, a_3) = \frac{a_2}{a_2 + a_3}$ we find that $\partial y/\partial a_2 > 0$ and $\partial y/\partial a_3 < 0$. This allows us to find the end points of the interval defining the α-cuts of the fuzzy conditional probability. We obtain

$$\overline{P}(A|B)[\alpha] = [0.5 + 0.25\alpha, 1 - 0.25\alpha], \tag{3.58}$$

for all $\alpha \in [0,1]$. Let $\overline{a}_i[\alpha] = [a_{i1}(\alpha), a_{i2}(\alpha)]$ for $1 \leq i \leq 4$. We see that; (1) $a_{21}(\alpha)$ and $a_{32}(\alpha)$ are feasible; and (2) $a_{22}(\alpha)$ and $a_{31}(\alpha)$ are feasible. What we did, to obtain the left end point of the interval in equation (3.58), was in the function f we : (1) substituted the left end point of the interval $\overline{a}_2[\alpha]$, which is $0.2 + 0.1\alpha$, for a_2; and (2) substituted the right end point of the interval $\overline{a}_3[\alpha]$, which is $0.2 - 0.1\alpha$, for a_3. To get the right end point of the interval: (1) substitute the right end point of the interval $\overline{a}_2[\alpha]$ for a_2; and (2) substitute the left end point of $\overline{a}_3[\alpha]$ for a_3. Hence, $\overline{P}(A|B) = (0.5/0.75/1)$ a triangular fuzzy number.

We will use the first definition of fuzzy conditional probability in the remainder of this book. Now we will show the basic properties of fuzzy conditional probability which are:

1. $0 \leq \overline{P}(A|B) \leq 1$;

2. $\overline{P}(B|B) = 1$;

3. $\overline{P}(A_1 \cup A_2|B) \leq \overline{P}(A_1|B) + \overline{P}(A_2|B)$, if $A_1 \cap A_2 = \phi$;

4. $\overline{P}(A|B) = 1$, if $B \subseteq A$; and

5. $\overline{P}(A|B) = 0$, if $B \cap A = \phi$.

Items (1), (2) and (4) follow immediately from our definition of fuzzy conditional probability. If we define the value of an empty sum to be zero, then (5) is true since the numerator in equation (3.55) will be empty when A and B are disjoint. So let us only present an argument for property (3). We show that $\overline{P}(A_1 \cup A_2|B)[\alpha] \subseteq \overline{P}(A_1|B)[\alpha] + \overline{P}(A_2|B)[\alpha]$, for $0 \leq \alpha \leq 1$. Choose and fix α. Let x belong to $\overline{P}(A_1 \cup A_2|B)[\alpha]$. Then $x = \frac{r+s}{t}$ where: (1) r is the sum of the a_i for $x_i \in A_1 \cap B$; (2) s is the sum of the a_i for $x_i \in A_2 \cap B$; and (3) t is the sum of the a_i for $x_i \in B$. Of course, the $a_i \in \overline{a}_i[\alpha]$ all i and the sum of the a_i equals one. Then $r/t \in \overline{P}(A_1|B)[\alpha]$ and $s/t \in \overline{P}(A_2|B)[\alpha]$. Hence $x \in \overline{P}(A_1|B)[\alpha] + \overline{P}(A_2|B)[\alpha]$ and the result follows. The following example shows that we may obtain $\overline{P}(A_1 \cup A_2|B) \neq \overline{P}(A_1|B) + \overline{P}(A_2|B)$ when $A_1 \cap A_2 = \phi$.

Example 3.6.3

We will use the same data as in Examples 3.6.1 and 3.6.2 but let $A_1 = \{x_2\}$ and $A_2 = \{x_3\}$ so that $A_1 \cup A_2 = B$ and $\overline{P}(A_1 \cup A_2|B) = 1$. We quickly determine, as in Example 3.6.2 that $\overline{P}(A_1|B) = (0.5/0.75/1)$ and $\overline{P}(A_2|B) = (0/0.25/0.5)$ so that $\overline{P}(A_1|B) + \overline{P}(A_2|B) = (0.5/1/1.5)$. Notice that when we add two fuzzy numbers both in $[0, 1]$ we can get a fuzzy number not contained in $[0, 1]$. Clearly, $\overline{P}(A_1 \cup A_2|B) \leq \overline{P}(A_1|B) + \overline{P}(A_2|B)$ but they are not equal.

3.7   Fuzzy Independence

We will present two definitions of two events A and B being independent and two versions of the first definition. We then argue against the second definition and therefore adopt the first definition for this book.

The first definition uses fuzzy conditional probability from the previous section. We will say that A and B are strongly independent if

$$\overline{P}(A|B) = \overline{P}(A), \tag{3.59}$$

and

$$\overline{P}(B|A) = \overline{P}(B). \tag{3.60}$$

Since the equality in equations (3.59) and (3.60) are sometimes difficult to satisfy we also have a weaker definition of independence. We will say that A and B are weakly independent if

$$\overline{P}(A|B)[1] = \overline{P}(A)[1], \tag{3.61}$$

and

$$\overline{P}(B|A)[1] = \overline{P}(B)[1]. \tag{3.62}$$

In the weaker version of independence we only require the equality for the core (where the membership values are one) of the fuzzy numbers. Clearly, if they are strongly independent they are weakly independent.

Our second definition of independence follows from the usual way of specifying independence in the crisp (non-fuzzy) case. We say that A and B are independent if

$$\overline{P}(A \cap B) = \overline{P}(A)\overline{P}(B). \tag{3.63}$$

We now argue in favor of the first definition because, due to fuzzy multiplication, it would be very rare to have the equality in equation (3.63) hold. The following example shows that events can be strongly independent (first definition) but it is too difficult for them to be independent by the second definition.

Example 3.7.1

Let $n = 4$ and $\bar{a}_i = (0.2/0.25/0.3)$, $1 \leq i \leq 4$. Also let $A = \{x_1, x_2\}$ and $B = \{x_2, x_3\}$. First, we easily see that $\overline{P}(A) = (0.4/0.5/0.6) = \overline{P}(B)$. To find $\overline{P}(A|B)$ we need to compute

$$\overline{P}(A|B)[\alpha] = \{\frac{a_2}{a_2 + a_3} | \quad \mathbf{S} \quad \}, \tag{3.64}$$

for all α. We do this as in Example 3.6.2 and obtain $\overline{P}(A|B) = (0.4/0.5/0.6) = \overline{P}(A)$. Similarly we see that $\overline{P}(B|A) = (0.4/0.5/0.6) = \overline{P}(B)$ and A and B are strongly independent. Now we go to the second definition and find $\overline{P}(A \cap B) = \bar{a}_2 = (0.2/0.25/0.3)$. But $\overline{P}(A)\overline{P}(B) \approx (0.16/0.25/0.36)$ a triangular shaped fuzzy number. Even if $\overline{P}(A|B)$, $\overline{P}(A)$ and $\overline{P}(B)$ are all triangular fuzzy numbers, $\overline{P}(A)\overline{P}(B)$ will not be a triangular fuzzy number. Because of fuzzy multiplication we would not expect the second definition of independence to hold, except in rare cases.

We will adopt the first definition of independence. Now let us see what are the basic properties of independence for fuzzy probabilities. In crisp probability theory we know that if A and B are independent so are A and B', and so are A' and B, and so are A' and B', where the "prime" denotes complement. However, this result may or may not be correct for strong independence and fuzzy probabilities. For the data in Example 3.7.1, A and B are strongly independent and so are A and B', as are A' and B, and this is also true for A' and B'. The following example shows that this is not always true.

Example 3.7.2

Let $n = 4$ but now $\bar{a}_i = (0.2/0.25/0.30)$, $1 \leq i \leq 3$ with $\bar{a}_4 = (0.1/0.25/0.4)$, and $A = \{x_1, x_2\}$, $B = \{x_2, x_3\}$. As in Example 3.7.1 we find that A and B are strongly independent but we can now show that A and B' are not strongly independent. That is, we argue that $\overline{P}(A|B') \neq \overline{P}(A)$. We know that $\overline{P}(A) = (0.4/0.5/0.6)$. To find α-cuts of $\overline{P}(A|B')$ we compute

$$\{\frac{a_1}{a_1 + a_4} | \quad \mathbf{S} \quad \}. \tag{3.65}$$

Now the fraction in equation (3.65) is an increasing function of a_1 but it is decreasing in a_4. Hence we obtain

$$\overline{P}(A|B')[\alpha] = [\frac{0.2 + 0.05\alpha}{0.6 - 0.1\alpha}, \frac{0.3 - 0.05\alpha}{0.4 + 0.1\alpha}], \tag{3.66}$$

for all $\alpha \in [0, 1]$. So $\overline{P}(A|B') \approx (\frac{1}{3}/0.5/0.75)$ a triangular shaped fuzzy number. We see that A and B' are not strongly independent.

However, the situation is possibly changed for weakly independent. Suppose all the $\bar{a}_i$ are triangular fuzzy numbers, A and B are not disjoint, $\bar{p}_i[1] = \frac{1}{n}$, $1 \leq i \leq n$, and A and B are weakly independent. Then it is true that: (1) A and B' are weakly independent; (2) A' and B are weakly independent; and (3) A' and B' are also weakly independent?

3.8   Fuzzy Bayes' Formula

Let A_i, $1 \leq i \leq m$, be a partition of $X = \{x_1, ..., x_n\}$. That is, the A_i are non-empty, mutually disjoint and their union is X. We do not know the probability of the A_i but we do know the conditional probability of A_i given the state of nature. There is a finite set of chance events, also called the states of nature, $\mathcal{S} = \{S_1, ..., S_K\}$ over which we have no control. What we do know is

$$a_{ik} = P(A_i|S_k), \tag{3.67}$$

for $1 \leq i \leq m$ and $1 \leq k \leq K$. If the operative state of nature is S_k, then the a_{ik} give the probabilities of the events A_i.

We do not know the probabilities of the states of nature, so we enlist a group of experts to give their estimates of $a_k = P(S_k)$. The a_k, $1 \leq k \leq K$, is called the prior probability distribution over the states of nature. From their estimates we construct fuzzy probabilities $\bar{a}_k$, see Section 3.3. We first present Bayes' formula using the crisp probabilities for the states of nature.

The probability that the state of nature S_k is in force, given the information that outcome A_j has occurred, is given by Bayes' formula

$$P(S_k|A_j) = \frac{P(A_j|S_k)P(S_k)}{\sum_{k=1}^{K} P(A_j|S_k)P(S_k)}, \tag{3.68}$$

for $1 \leq k \leq K$. The $a_{kj} = P(S_k|A_j)$, $1 \leq k \leq K$, is the posterior probability distribution over the states of nature.

Let us see how this result may be used. Using the a_{ik} and the prior distribution a_k, we may calculate $P(A_i)$ as follows

$$P(A_i) = \sum_{k=1}^{K} P(A_i|S_k)P(S_k), \tag{3.69}$$

for $1 \leq i \leq m$. Now we gather some information and observe that event A_j has occurred. We update the prior to the posterior and then obtain improved estimates of the probabilities for the A_i as

$$P(A_i) = \sum_{k=1}^{K} P(A_i|S_k)P(S_k|A_j), \tag{3.70}$$

for $1 \leq i \leq m$.

Now substitute $\overline{a}_k$ for a_k, $1 \le k \le K$. Suppose we observe that event A_j has occurred. Alpha-cuts of the fuzzy posterior distribution are

$$\overline{P}(S_k|A_j)[\alpha] = \{\frac{a_{jk}a_k}{\sum_{k=1}^{K} a_{jk}a_k} \mid \mathbf{S}\ \},\qquad (3.71)$$

for $1 \le k \le K$, where $\mathbf{S}$ is the statement " $a_k \in \overline{a}_k[\alpha]$, $1 \le k \le K$, $\sum_{k=1}^{K} a_k = 1$ ". It may not be difficult to find these α-cuts. Suppose $K = 3$, k=2 and let

$$f(a_1, a_2, a_3) = \frac{a_{j2}a_2}{\sum_{k=1}^{3} a_{jk}a_k}.\qquad (3.72)$$

Then $\partial f/\partial a_1 < 0$, $\partial f/\partial a_2 > 0$ and $\partial f/\partial a_3 < 0$. Let $\overline{a}_k[\alpha] = [a_{k1}(\alpha), a_{k2}(\alpha)]$ for $k = 1, 2, 3$. If $a_{12}(\alpha) + a_{21}(\alpha) + a_{32}(\alpha) = 1$ and $a_{11}(\alpha) + a_{22}(\alpha) + a_{31}(\alpha) = 1$, then we get the left (right) end point of the interval for the α-cut by substituting $a_{12}(\alpha)$, $a_{21}(\alpha)$, $a_{32}(\alpha)$ $(a_{11}(\alpha)$, $a_{22}(\alpha)$, $a_{31}(\alpha))$ for a_1, a_2, a_3, respectively.

Once we have the fuzzy posterior, we may update our fuzzy probability for the A_i.

There is an alternate method of computing fuzzy Bayes' rule. We could just substitute the fuzzy numbers $\overline{P}(S_k) = \overline{a}_k$ into equation (3.68) and compute the result. However, we would come up against the same problem noted in Section 3.6, see Example 3.6.1, where the result can produce a fuzzy number not in the interval $[0, 1]$.

3.9 Applications

We first present two applications of fuzzy probability followed by two applications of fuzzy conditional probability. Then we give an application of fuzzy Bayes' formula. We have a change of notation in this section: we will use p_i for probability values instead of a_i used in previous sections of this chapter.

3.9.1 Blood Types

There are four basic blood types: A, B, AB and O. A certain city is going to have a blood drive and they want to know that if they select one person at random, from the pool of possible blood donors, what is the probability that this person does not have blood type O? Type O is the universal blood donor group. They conduct a random sample of 1000 people from the pool of blood donors and they determine the following point estimates: (1) $p_a = 0.33$, or 33% have blood type A; (2) $p_b = 0.23$, or 23% have blood type B; (3) $p_{ab} = 0.35$, or 35% are of blood type AB; and (4) $p_o = 0.09$, or 9% belong to blood type O. Because these are point estimates based on a random sample we will substitute fuzzy numbers for these probabilities. Let $\overline{p}_a = (0.3/0.33/0.36)$, $\overline{p}_b = (0.2/0.23/0.26)$, $\overline{p}_{ab} = (0.32/0.35/0.38)$ and $\overline{p}_o = (0.06/0.09/0.12)$. Next let $\overline{P}(O')$ stand for the fuzzy probability of a donor not having blood

α	$\overline{P}(O')[\alpha]$
0	[0.88,0.94]
0.2	[0.886,0.934]
0.4	[0.892,0.928]
0.6	[0.898,0.922]
0.8	[0.904,0.916]
1.0	0.91

Table 3.1: Alpha-Cuts of $\overline{P}(O')$

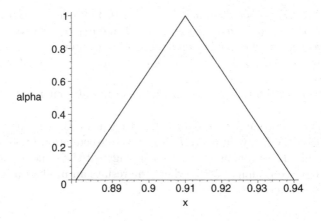

Figure 3.3: Fuzzy Probability in the Blood Type Application

type O. From the discussion in Section 3.5 we do not expect $\overline{P}(O')$ to equal $1 - \overline{P}(O)$. However, we find α-cuts of this fuzzy probability as

$$\overline{P}(O')[\alpha] = \{p_a + p_b + p_{ab}| \quad \textbf{S} \quad \}, \tag{3.73}$$

for all α, where $\textbf{S}$ denotes the statement "$p_a \in \overline{p}_a[\alpha]$, $p_b \in \overline{p}_b[\alpha]$, $p_{ab} \in \overline{p}_{ab}[\alpha]$, $p_o \in \overline{p}_o[\alpha]$, $p_a + p_b + p_{ab} + p_o = 1$". Define $\overline{p}_w[\alpha] = [p_{w1}(\alpha), p_{w2}(\alpha)]$ for $w = a, b, ab, o$. Let $\overline{P}(O')[\alpha] = [o_{n1}(\alpha), o_{n2}(\alpha)]$. We would like to substitute $p_{a1}(\alpha) + p_{b1}(\alpha) + p_{ab1}(\alpha)$ for $p_a + p_b + p_{ab}$ in equation (3.73) to get $o_{n1}(\alpha)$, but this set of $p'_i s$ is not feasible. What we mean is that $p_{a1}(0) + p_{b1}(0) + p_{ab1}(0) = 0.82$ and there is no value of $p_0 \in \overline{p}_o[0]$ that can make the sum equal to one. Therefore we need to use some numerical method to compute these α-cuts. We used Maple [12]. The Maple commands used to solve this problem are similar to those for Example 3.4.3.4 whose Maple commands are in Chapter 30. The results are displayed in Table 3.1. From the data in Table 3.1 we compute $\overline{P}(O')[\alpha] = [0.88 + 0.03\alpha, 0.94 - 0.03\alpha]$, or it is a triangular fuzzy number $(0.88/0.91/0.94)$, which is shown in Figure 3.3. See also Example 3.4.3.5.

3.9.2 Resistance to Surveys

Pollsters are concerned about the increased level of resistance of people to answer questions during surveys. They conduct a study using a random sample of n people, from a large population of people who are candidates for surveys, ages 18 to 40. This random sample is broken up into two subpopulations: (1) a population of n_1 "young" people with ages 18 to 29; and (2) a population n_2 ($n_1 + n_2 = n$) of "older" people ages 30-40. From this data they obtain the following point estimates: (1) $p_1 = 0.18$, or 18% of young people said they will respond to questions in a survey; (2) $p_2 = 0.05$, or 5% of the young people said that they would not respond to questions in a survey; (3) $p_3 = 0.68$, or 68% of the older people responded that they would participate in a survey; and (4) $p_4 = 0.09$, or 9% of the older people indicated that they would not participate in a survey. The pollsters want to know if they choose at random one person from this group, aged 18 to 40, what is the probability that this person is young or is someone who would refuse to respond to the questions in a survey?

All these probabilities were estimated and so we substitute fuzzy numbers for the p_i. Assume that we decide to use : (1) $\overline{p}_1 = (0.16/0.18/0.20)$ for p_1; (2) $\overline{p}_2 = (0.03/0.05/0.07)$ for p_2; (3) $\overline{p}_3 = (0.61/0.68/0.75)$ for p_3; and (4) $\overline{p}_4 = (0.07/0.09/0.11)$ for p_4. Let A be the event that the person is young and set B to be the event that the person will not respond to a survey. We therefore wish to find the fuzzy probability $\overline{P}(A \cup B)$. From Section 3.5 we know that this fuzzy probability may not equal $\overline{P}(A) + \overline{P}(B) - \overline{P}(A \cap B)$, since A and B are not disjoint. However, we may still find the α-cuts of $\overline{P}(A \cup B)$ as follows

$$\overline{P}(A \cup B)[\alpha] = \{p_1 + p_2 + p_4| \quad \mathbf{S} \quad \}, \tag{3.74}$$

for $0 \leq \alpha \leq 1$, where $\mathbf{S}$ is the statement " $p_i \in \overline{p}_i[\alpha]$, $1 \leq i \leq 4$, and $p_1 + ... + p_4 = 1$". Let $\overline{p}_i[\alpha] = [p_{i1}(\alpha), p_{i2}(\alpha)]$, $1 \leq i \leq 4$ and set $\overline{P}(A \cup B)[\alpha] = [P_1(\alpha), P_2(\alpha)]$. Then

$$\overline{P}_1(\alpha) = p_{11}(\alpha) + p_{21}(\alpha) + p_{41}(\alpha), \tag{3.75}$$

and

$$\overline{P}_2(\alpha) = p_{12}(\alpha) + p_{22}(\alpha) + p_{42}(\alpha), \tag{3.76}$$

for all α because now these p_i's are feasible. What this means is: (1) for all α there is a $p_3 \in \overline{p}_3[\alpha]$ so that $p_{11}(\alpha) + p_{21}(\alpha) + p_{41}(\alpha) + p_3 = 1$; and (2) for all α there is a value of $p_3 \in \overline{p}_3[\alpha]$ so that $p_{12}(\alpha) + p_{22}(\alpha) + p_{42}(\alpha) + p_3 = 1$. The graph of the fuzzy probability $\overline{P}(A \cup B)$ is in Figure 3.4. It turns out that $\overline{P}(A \cup B)$ is a triangular fuzzy number $(0.26/0.32/0.38)$.

3.9.3 Testing for HIV

It is important to have an accurate test for the HIV virus. Suppose we have a test, which we shall call $\mathcal{T}$, and we want to see how accurately it predicts that

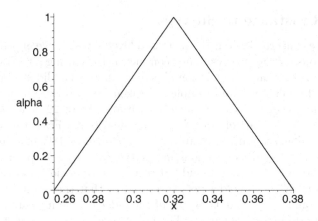

Figure 3.4: Fuzzy Probability in the Survey Application

a person has the HIV virus. Let A_1 be the event that a person is infected with the HIV virus, and A_2 is the event that a person is not infected. Also, let B_1 be the event that the test $\mathcal{T}$ is "positive" indicating that the person has the virus and set B_2 to be the event that the test $\mathcal{T}$ gives the result of "negative", or the person does not have the virus. We want to find the conditional probability of A_1 given B_1, or $P(A_1|B_1)$. To estimate this probability we gather some data. From a large population of the "at-risk" population we take a random sample to estimate the probabilities $p_{11} = P(A_1 \cap B_1)$, $p_{12} = P(A_1 \cap B_2)$, $p_{21} = P(A_2 \cap B_1)$ and $p_{22} = P(A_2 \cap B_2)$. Assume we obtain the estimates $p_{11} = 0.095$, $p_{12} = 0.005$, $p_{21} = 0.045$ and $p_{22} = 0.855$. To show the uncertainty in these point estimates we now substitute fuzzy numbers for the p_{ij}. Let $\overline{p}_{11} = (0.092/0.095/0.098)$, $\overline{p}_{12} = (0.002/0.005/0.008)$, $\overline{p}_{21} = (0.042/0.045/0.048)$ and $\overline{p}_{22} = (0.825/0.855/0.885)$.
    The fuzzy probability we want is $\overline{P}(A_1|B_1)$ whose α-cuts are

$$\overline{P}(A_1|B_1)[\alpha] = \{\frac{p_{11}}{p_{11} + p_{21}}|\ \mathbf{S}\ \}, \tag{3.77}$$

for all α, where $\mathbf{S}$ is "$p_{ij} \in \overline{p}_{ij}[\alpha]$, $1 \le i, j \le 2$ and $p_{11} + ... + p_{22} = 1$". Let $H(p_{11}, p_{21}) = \frac{p_{11}}{p_{11}+p_{21}}$. We determine that H is an increasing function of p_{11} and it is a decreasing function of p_{21}. Hence, if $\overline{P}(A_1|B_1)[\alpha] = [P_l(\alpha), P_u(\alpha)]$, then

$$P_l(\alpha) = H(p_{111}(\alpha), p_{212}(\alpha)), \tag{3.78}$$

and

$$P_u(\alpha) = H(p_{112}(\alpha), p_{211}(\alpha)], \tag{3.79}$$

where $\overline{p}_{ij}[\alpha] = [p_{ij1}(\alpha), p_{ij2}(\alpha)]$, for $1 \le i, j \le 2$. Equations (3.78) and (3.79) are correct because : (1) for any α there are values of $p_{12} \in \overline{p}_{12}[\alpha]$ and $p_{22} \in \overline{p}_{22}[\alpha]$ so that $p_{111}(\alpha) + p_{212}(\alpha) + p_{12} + p_{22} = 1$; and (2) for any α there are

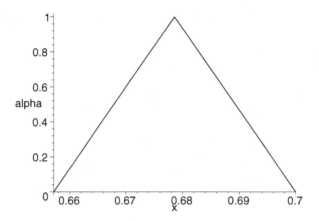

Figure 3.5: Fuzzy Probability of HIV Given Test Positive

values of $p_{12} \in \overline{p}_{12}[\alpha]$ and $p_{22} \in \overline{p}_{22}[\alpha]$ so that $p_{112}(\alpha) + p_{211}(\alpha) + p_{12} + p_{22} = 1$. The graph of $\overline{P}(A_1|B_1)$ is in Figure 3.5. We get a triangular fuzzy number $(\frac{92}{140}/\frac{95}{140}/\frac{98}{140})$ for $\overline{P}(A_1|B_1)$.

3.9.4 Color Blindness

Some people believe that red-green color blindness is more prevalent in males than in females. To test this hypothesis we gather a random sample form the adult US population. Let M be the event a person is male, F is the event the person is female, C is the event the person has red-green color blindness and C' is the event he/she does not have red-green color blindness. From the data we obtain point estimates of the following probabilities: (1) $p_{11} = P(M \cap C) = 0.040$; (2) $p_{12} = P(M \cap C') = 0.493$; (3) $p_{21} = P(F \cap C) = 0.008$; and (4) $p_{22} = P(F \cap C') = 0.459$. The uncertainty in these point estimates will be shown in their fuzzy values : (1) $\overline{p}_{11} = (0.02/0.04/0.06)$; (2) $\overline{p}_{12} = (0.463/0.493/0.523)$; (3) $\overline{p}_{21} = (0.005/0.008/0.011)$; and (4) $\overline{p}_{22} = (0.439/0.459/0.479)$.

We wish to calculate the fuzzy conditional probabilities $\overline{P}(M|C)$ and $\overline{P}(F|C)$. The α-cuts of the first fuzzy probability are

$$\overline{P}(M|C)[\alpha] = \{\frac{p_{11}}{p_{21} + p_{11}}| \text{ } \textbf{S} \text{ } \}, \tag{3.80}$$

for $\alpha \in [0,1]$ and $\textbf{S}$ denotes "$p_{ij} \in p_{ij}[\alpha]$, $1 \leq i, j \leq 2$ and $p_{11} + ... + p_{22} = 1$". Let $H(p_{11}, p_{21}) = \frac{p_{11}}{p_{11}+p_{21}}$. Then H is an increasing function of p_{11} but decreasing in p_{21}. So, as in the previous application, we obtain

$$\overline{P}(M|C)[\alpha] = [H(p_{111}(\alpha), p_{212}(\alpha)), H(p_{112}(\alpha), p_{211}(\alpha))], \tag{3.81}$$

for all α. This fuzzy conditional probability is shown in Figure 3.6. Maple commands for this figure are in Chapter 30.

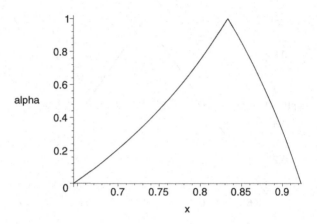

Figure 3.6: Fuzzy Probability of Male Given Color Blind

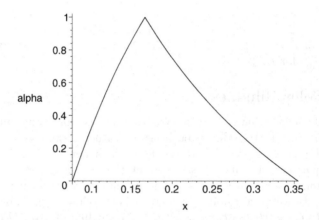

Figure 3.7: Fuzzy Probability of Female Given Color Blind

Alpha-cuts of the second fuzzy conditional probability are

$$\overline{P}(F|C)[\alpha] = \{\frac{p_{21}}{p_{21} + p_{11}}|\ \ \mathbf{S}\ \ \}, \qquad (3.82)$$

for all α. Let $G(p_{11}, p_{21}) = \frac{p_{21}}{p_{21}+p_{11}}$. Then G increases in p_{21} but decreases in p_{11}. We may check that the following result is "feasible".

$$\overline{P}(F|C)[\alpha] = [G(p_{112}(\alpha), p_{211}(\alpha)), G(p_{111}(\alpha), p_{212}(\alpha))], \qquad (3.83)$$

for all α. This fuzzy probability is in Figure 3.7.

Do we obtain $\overline{P}(F|C) \leq \overline{P}(M|C)$, where here $\leq$ means "less that or equal to"? See Section 2.5 of Chapter 2.

3.9.5 Fuzzy Bayes

This is a numerical application showing how a fuzzy probability $\overline{P}(A_1)$ changes from the fuzzy prior to the fuzzy posterior. Let there be only two states of nature S_1 and S_2 with fuzzy prior probabilities $\overline{p}_1 = \overline{P}(S_1) = (0.3/0.4/0.5)$ and $\overline{p}_2 = \overline{P}(S_2) = (0.5/0.6/0.7)$. There are also only two events A_1 and A_2 in the partition of X with known conditional probabilities $p_{11} = P(A_1|S_1) = 0.2$, $p_{21} = P(A_2|S_1) = 0.8$, $p_{12} = P(A_1|S_2) = 0.7$ and $p_{22} = P(A_2|S_2) = 0.3$. We first find the fuzzy probability $\overline{P}(A_1)$ using the fuzzy prior probabilities. Its α-cuts are

$$\overline{P}(A_1)[\alpha] = \{(0.2)p_1 + (0.7)p_2| \quad \mathbf{S} \quad \}, \tag{3.84}$$

where $\mathbf{S}$ is " $p_i \in \overline{p}_i[\alpha]$, $1 \leq i \leq 2$, and $p_1 + p_2 = 1$". We easily evaluate this equation (3.84) and get the triangular fuzzy number $\overline{P}(A_1) = (0.41/0.50/0.59)$.

Now suppose we have information that event A_1 will occur. We need to obtain $\overline{P}(S_1|A_1)$ and $\overline{P}(S_2|A_1)$ from fuzzy Bayes' formula in Section 3.8. We first compute

$$\overline{P}(S_1|A_1)[\alpha] = \{\frac{(0.2)p_1}{(0.2)p_1 + (0.7)p_2}| \quad \mathbf{S} \quad \}, \tag{3.85}$$

and

$$\overline{P}(S_2|A_1)[\alpha] = \{\frac{(0.7)p_2}{(0.2)p_1 + (0.7)p_2}| \quad \mathbf{S} \quad \} \tag{3.86}$$

where $\mathbf{S}$ is "$p_i \in \overline{p}_i[\alpha]$, $i = 1, 2$ and $p_1 + p_2 = 1$". Both α-cuts are easily found. Let $\overline{p}_i[\alpha] = [p_{i1}(\alpha), p_{i2}(\alpha)]$, for $i = 1, 2$. Then

$$\overline{P}(S_1|A_1)[\alpha] = [\frac{0.2p_{11}(\alpha)}{0.2p_{11}(\alpha) + 0.7p_{22}(\alpha)}, \frac{0.2p_{12}(\alpha)}{0.2p_{12}(\alpha) + 0.7p_{21}(\alpha)}], \tag{3.87}$$

and

$$\overline{P}(S_2|A_1)[\alpha] = [\frac{0.7p_{21}(\alpha)}{0.2p_{12}(\alpha) + 0.7p_{21}(\alpha)}, \frac{0.7p_{22}(\alpha)}{0.2p_{11}(\alpha) + p_{22}(\alpha)}], \tag{3.88}$$

for all α. We get

$$\overline{P}(S_1|A_1)[\alpha] = [\frac{0.06 + 0.02\alpha}{0.55 - 0.05\alpha}, \frac{0.10 - 0.02\alpha}{0.45 + 0.05\alpha}], \tag{3.89}$$

and

$$\overline{P}(S_2|A_1)[\alpha] = [\frac{0.35 + 0.07\alpha}{0.45 + 0.05\alpha}, \frac{0.49 - 0.07\alpha}{0.55 - 0.05\alpha}]. \tag{3.90}$$

Now we may compute $\overline{P}(A_1)$ using the fuzzy posterior probabilities. It has α-cuts

$$(0.2)\overline{P}(S_1|A_1)[\alpha] + (0.7)\overline{P}(S_2|A_1)[\alpha]. \tag{3.91}$$

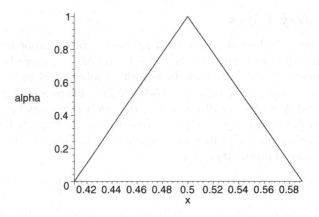

Figure 3.8: $\overline{P}(A_1)$ Using the Fuzzy Prior

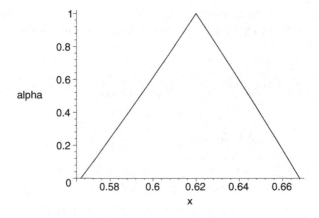

Figure 3.9: $\overline{P}(A_1)$ Using the Fuzzy Posterior

The graphs of $\overline{P}(A_1)$ using the fuzzy prior is in Figure 3.8 and $\overline{P}(A_1)$ from the fuzzy posterior is shown in Figure 3.9. Maple commands for Figure 3.9 are in Chapter 30. The fuzzy number in Figure 3.8 is the triangular fuzzy number (0.41/0.50/0.59). The fuzzy number in Figure 3.9 may look like a triangular fuzzy number, but it is not. The sides of the fuzzy number in Figure 3.9 are slightly curved and are not straight lines. Why does $\overline{P}(A_1)$ in Figure 3.9 lie to the right of $\overline{P}(A_1)$ shown in Figure 3.8?

3.10   References

1. J.J. Buckley: Fuzzy Probabilities: New Approach and Applications, Physica-Verlag, Heidelberg, Germany, 2003.

2. J.J. Buckley: Fuzzy Probabilities and Fuzzy Sets for Web Planning, Springer, Heidelberg, Germany, 2004.

3. J.J. Buckley and E. Eslami: An Introduction to Fuzzy Logic and Fuzzy Sets, Springer-Verlag, Heidelberg, Germany, 2002.

4. Frontline Systems (www.frontsys.com).

5. R.V. Hogg and E.A. Tanis: Probability and Statistical Inference, Sixth Edition, Prentice Hall, Upper Saddle River, N.J., 2001.

6. G.J. Klir: Fuzzy Arithmetic with Requisite Constraints, Fuzzy Sets and Systems, 91(1997)147-161.

7. G.J. Klir: The Role of Constrained Fuzzy Arithmetic in Engineering, In: B.M. Ayyub and M.M. Gupta (eds.), Uncertainty Analysis in Engineering and Science: Fuzzy Logic, Statistics, and Neural Network Approach, Kluwer, Dordrecht, 1998, 1-19.

8. G.J. Klir and J.A. Cooper: On Constrained Fuzzy Arithmetic, Proc. 5th Int. IEEE Conf. on Fuzzy Systems, New Orleans, 1996, 1285-1290.

9. G.J. Klir and Y. Pan: Constrained Fuzzy Arithmetic: Basic Questions and Some Answers, Soft Computing, 2(1998)100-108.

10. G.J. Klir and B. Yuan: Fuzzy Sets and Fuzzy Logic: Theory and Applications, Prentice Hall, Upper Saddle River, N.J., 1995.

11. V. Kreinovich, H.T. Nguyen, S. Ferson and L. Ginzburg: From Computation with Guaranteed Intervals to Computation with Confidence Intervals: A New Application of Fuzzy Techniques, Proc. 21st NAFIPS 2002, New Orleans, LA, June 27-29, 2002, 418-422.

12. Maple 9, Waterloo Maple Inc. Waterloo, Canada.

13. M. Navara and Z. Zabokrtsky: How to Make Constrained Fuzzy Arithmetic Efficient, Soft Computing, 6(2001)412-417.

14. Y. Pan and G.J. Klir: Bayesian Inference Based on Interval-Valued Prior Distributions and Likelihoods, J. of Intelligent and Fuzzy Systems, 5(1997)193-203.

15. Y. Pan and B. Yuan: Baysian Inference of Fuzzy Probabilities, Int. J. General Systems, 26(1997)73-90.

16. Pim van den Broek, University of Twente, The Netherlands (p.m.vandenbroek@ewi.utwente.nl).

17. www.solver.com.

18. H.A. Taha: Operations Research, Fifth Edition, Macmillan, N.Y., 1992.

Chapter 4

Discrete Fuzzy Random Variables

4.1 Introduction

We start with the fuzzy binomial. The crisp binomial probability mass function, usually written $b(m, p)$ where m is the number of independent experiments and p is the probability of a "success" in each experiment, has one parameter p. We assume that p is not known exactly and is to be estimated from a random sample or from expert opinion. In either case the result is, justified in Sections 3.2 and 3.3, that we substitute a fuzzy number $\overline{p}$ for p to get the fuzzy binomial. Then we discuss the fuzzy Poisson probability mass function. The crisp Poisson probability mass function has one parameter, usually written λ, which we also assume is not known exactly. A fuzzy estimator for λ, derived from data, is in Chapter 10. Hence we substitute fuzzy number $\overline{\lambda}$ for λ to obtain the fuzzy Poisson probability mass function. The fuzzy binomial and the fuzzy Poisson comprises the next two sections. We look at some applications of these two discrete fuzzy probability mass functions in Section 4.4.

4.2 Fuzzy Binomial

As before $X = \{x_1, ..., x_n\}$ and let E be a non-empty, proper, subset of X. We have an experiment where the result is considered a "success" if the outcome x_i is in E. Otherwise, the result is considered a "failure". Let $P(E) = p$ so that $P(E') = q = 1 - p$. $P(E)$ is the probability of success and $P(E')$ is the probability of failure. We assume that $0 < p < 1$.

James J. Buckley: *Fuzzy Probability and Statistics*, StudFuzz **196**, 51–60 (2006)
www.springerlink.com © Springer-Verlag Berlin Heidelberg 2006

Suppose we have m independent repetitions of this experiment. If $P(r)$ is the probability of r successes in the m experiments, then

$$P(r) = \binom{m}{r} p^r q^{m-r}, \tag{4.1}$$

for $r = 0, 1, 2, ..., m$, gives the binomial distribution.

In these experiments let us assume that $P(E)$ is not known precisely and it needs to be estimated, or obtained from expert opinion. So the p value is uncertain and we substitute $\bar{p}$ for p and a $\bar{q}$ for q so that there is a $p \in \bar{p}[1]$ and a $q \in \bar{q}[1]$ with $p + q = 1$. $\bar{q}$ could equal $1 - \bar{p}$. Now let $\overline{P}(r)$ be the fuzzy probability of r successes in m independent trials of the experiment. Under our restricted fuzzy arithmetic we obtain

$$\overline{P}(r)[\alpha] = \{\binom{m}{r} p^r q^{m-r} | \ \mathbf{S} \ \}, \tag{4.2}$$

for $0 \leq \alpha \leq 1$, where $\mathbf{S}$ is the statement "$p \in \bar{p}[\alpha], q \in \bar{q}[\alpha]$, $p + q = 1$". Notice that $\overline{P}(r)$ is not $\binom{m}{r} \bar{p}^r \bar{q}^{m-r}$. If $\overline{P}(r)[\alpha] = [P_{r1}(\alpha), P_{r2}(\alpha)]$, then

$$P_{r1}(\alpha) = min\{\binom{m}{r} p^r q^{m-r} | \ \mathbf{S} \ \}, \tag{4.3}$$

and

$$P_{r2}(\alpha) = max\{\binom{m}{r} p^r q^{m-r} | \ \mathbf{S} \ \}. \tag{4.4}$$

Example 4.2.1

Let $p = 0.4$, $q = 0.6$ and $m = 3$. Since p and q are uncertain we use $\bar{p} = (0.3/0.4/0.5)$ for p and $\bar{q} = (0.5/0.6/0.7)$ for q. Now we will calculate the fuzzy number $\overline{P}(2)$. If $p \in \bar{p}[\alpha]$ then $q = 1 - p \in \bar{q}[\alpha]$. Equations (4.3) and (4.4) become

$$P_{r1}(\alpha) = min\{3p^2q| \ \mathbf{S} \ \}, \tag{4.5}$$

and

$$P_{r2}(\alpha) = max\{3p^2q| \ \mathbf{S} \ \}. \tag{4.6}$$

Since $d(3p^2(1-p))/dp > 0$ on $\bar{p}[0]$ we obtain

$$\overline{P}(2)[\alpha] = [3(p_1(\alpha))^2(1 - p_1(\alpha)), 3(p_2(\alpha))^2(1 - p_2(\alpha))], \tag{4.7}$$

where $\bar{p}[\alpha] = [p_1(\alpha), p_2(\alpha)] = [0.3 + 0.1\alpha, 0.5 - 0.1\alpha]$.

Alpha-cuts of the fuzzy mean and the fuzzy variance of the fuzzy binomial distribution are calculated as in equations (3.51) and (3.52) in Chapter 3, respectively. In the crisp case we know $\mu = mp$ and $\sigma^2 = mpq$. Does $\bar{\mu} = m\bar{p}$

and $\overline{\sigma}^2 = m\overline{p} \cdot \overline{q}$? We now argue that the correct result is $\overline{\mu} \leq m\overline{p}$ and $\overline{\sigma}^2 \leq m\overline{p} \cdot \overline{q}$. We see that

$$\overline{\mu}[\alpha] = \{\sum_{r=1}^{m} r\binom{m}{r} p^r q^{m-r}| \quad \mathbf{S} \quad \}, \tag{4.8}$$

which simplifies to

$$\overline{\mu}[\alpha] = \{mp| \quad \mathbf{S} \quad \}. \tag{4.9}$$

Let $s \in \overline{\mu}[\alpha]$. Then $s = mp$ for $p \in \overline{p}[\alpha]$, $q \in \overline{q}[\alpha]$ and $p + q = 1$. Hence, $s \in m\overline{p}[\alpha]$. So $\overline{\mu} \leq m\overline{p}$. To show they may not be equal let $\overline{p} = (0.2/0.3/0.4)$ and $\overline{q} = (0.65/0.7/0.75)$. Then $\overline{\mu}[0] = m[0.25, 0.35]$ but $m\overline{p}[0] = m[0.2, 0.4]$. If $\overline{q} = 1 - \overline{p}$, then $\overline{\mu} = m\overline{p}$.

To show $\overline{\sigma}^2 \leq m\overline{p} \cdot \overline{q}$ we see first, as we get equation (4.9) from equation (4.8), that

$$\overline{\sigma}^2[\alpha] = \{mpq| \quad \mathbf{S} \quad \}. \tag{4.10}$$

Then we argue, just like before, that given $s \in \overline{\sigma}^2[\alpha]$, then $s \in m\overline{p}[\alpha]\overline{q}[\alpha]$. This shows $\overline{\sigma}^2 \leq m\overline{p} \cdot \overline{q}$. Now, to show that they may not be equal let $\overline{p}$ and $\overline{q}$ be given as above. Then $m\overline{p}[0]\overline{q}[0] = m[0.13, 0.30]$ but $\overline{\sigma}^2[0] = m[(0.25)(0.75), (0.35)(0.65)] = m[0.1875, 0.2275]$.

Example 4.2.2

We may find the α-cuts of $\overline{\sigma}^2$ if $\overline{q} = 1 - \overline{p}$. Let $\overline{p} = (0.4/0.6/0.8)$ and $\overline{q} = (0.2/0.4/0.6)$. Then

$$\overline{\sigma}^2[\alpha] = \{mp(1 - p)|p \in \overline{p}[\alpha]\}, \tag{4.11}$$

from equation (4.10). Let $h(p) = mp(1 - p)$. We see that $h(p)$: (1) is increasing on $[0, 0.5]$; (2) has its maximum of $0.25m$ at $p = 0.5$; and (3) is decreasing on $[0.5, 1]$. So, the evaluation of equation (4.11), see Section 3.4.3, depends if $p = 0.5$ belongs to the α-cut of $\overline{p}$. Let $\overline{p}[\alpha] = [p_1(\alpha), p_2(\alpha)] = [0.4 + 0.2\alpha, 0.8 - 0.2\alpha]$. So, $p = 0.5$ belongs to the α-cut of $\overline{p}$ only for $0 \leq \alpha \leq 0.5$. Then

$$\overline{\sigma}^2[\alpha] = [h(p_2(\alpha)), 0.25m], \tag{4.12}$$

for $0 \leq \alpha \leq 0.5$, and

$$\overline{\sigma}^2[\alpha] = [h(p_2(\alpha)), h(p_1(\alpha))], \tag{4.13}$$

for $0.5 \leq \alpha \leq 1$. We substitute in for $p_1(\alpha)$ and $p_2(\alpha)$ to finally obtain $\overline{\sigma}^2$ and its graph, for $m = 10$, is in Figure 4.1.

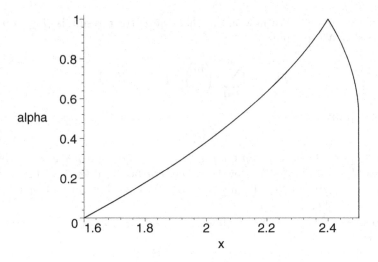

Figure 4.1: Fuzzy Variance in Example 4.2.2

4.3   Fuzzy Poisson

Let X be a random variable having the Poisson probability mass function. If $P(x)$ stands for the probability that $X = x$, then

$$P(x) = \frac{\lambda^x \exp(-\lambda)}{x!}, \qquad (4.14)$$

for $x = 0, 1, 2, 3, ...$, and parameter $\lambda > 0$. Now substitute fuzzy number $\overline{\lambda} > 0$ for λ to produce the fuzzy Poisson probability mass function. Set $\overline{P}(x)$ to be the fuzzy probability that $X = x$. Then we find α-cuts of this fuzzy number as

$$\overline{P}(x)[\alpha] = \{ \frac{\lambda^x \exp(-\lambda)}{x!} | \lambda \in \overline{\lambda}[\alpha] \}, \qquad (4.15)$$

for all $\alpha \in [0, 1]$. The evaluation of equation (4.15) depends on the relation of x to $\overline{\lambda}[0]$. Let $h(\lambda) = \frac{\lambda^x \exp(-\lambda)}{x!}$ for fixed x and $\lambda > 0$. We see that $h(\lambda)$ is an increasing function of λ for $\lambda < x$, the maximum value of $h(\lambda)$ occurs at $\lambda = x$, and $h(\lambda)$ is a decreasing function of λ for $\lambda > x$. Let $\overline{\lambda}[\alpha] = [\lambda_1(\alpha), \lambda_2(\alpha)]$, for $0 \leq \alpha \leq 1$. Then we see that : (1) if $\lambda_2(0) < x$, then $\overline{P}(x)[\alpha] = [h(\lambda_1(\alpha)), h(\lambda_2(\alpha))]$; and (2) if $x < \lambda_1(0)$, then $\overline{P}(x)[\alpha] = [h(\lambda_2(\alpha)), h(\lambda_1(\alpha))]$. The other case, where $x \in \overline{\lambda}[0]$, is explored in the following example.

Example 4.3.1

Let $x = 6$ and $\overline{\lambda} = (3/5/7)$. We see that $x \in [3, 7] = \overline{\lambda}[0]$. We determine $\overline{\lambda}[\alpha] = [3 + 2\alpha, 7 - 2\alpha]$. Define $\overline{P}(6)[\alpha] = [p_1(\alpha), p_2(\alpha)]$. To determine the

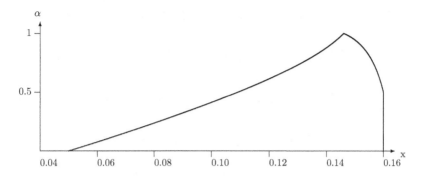

Figure 4.2: Fuzzy Probability in Example 4.3.1

α-cuts of $\overline{P}(6)$ we need to solve (see equations (2.24) and (2.25) in Chapter 2)

$$p_1(\alpha) = min\{h(\lambda)|\lambda \in \overline{\lambda}[\alpha]\}, \tag{4.16}$$

and

$$p_2(\alpha) = max\{h(\lambda)|\lambda \in \overline{\lambda}[\alpha]\}. \tag{4.17}$$

It is not difficult to solve equations (4.16) and (4.17) producing

$$\overline{P}(6)[\alpha] = [h(3 + 2\alpha), h(6)], \tag{4.18}$$

for $0 \le \alpha \le 0.5$, and

$$\overline{P}(6)[\alpha] = [h(3 + 2\alpha), h(7 - 2\alpha)], \tag{4.19}$$

for $0.5 \le \alpha \le 1$. The graph of $\overline{P}(6)$ is shown in Figure 4.2.

   Let us consider another, slightly more complicated, example of finding fuzzy probabilities using the fuzzy Poisson.

Example 4.3.2

Let $\overline{\lambda} = (8/9/10)$ and define $\overline{P}([3, \infty))$ to be the fuzzy probability that $X \ge 3$. Also let $\overline{P}([3, \infty))[\alpha] = [q_1(\alpha), q_2(\alpha)]$. Then

$$q_1(\alpha) = min\{1 - \sum_{x=0}^{2} h(\lambda)|\lambda \in \overline{\lambda}[\alpha]\}, \tag{4.20}$$

and

$$q_2(\alpha) = max\{1 - \sum_{x=0}^{2} h(\lambda)|\lambda \in \overline{\lambda}[\alpha]\}, \tag{4.21}$$

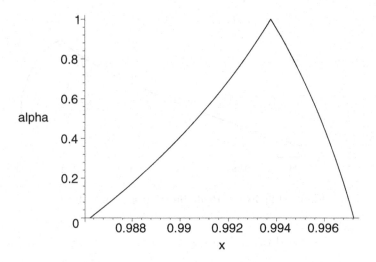

Figure 4.3: Fuzzy Probability in Example 4.3.2

for all α. Let $k(\lambda) = 1 - [\sum_{x=0}^{2} h(\lambda)]$. Then $dk/d\lambda > 0$ for $\lambda > 0$ Hence, we may evaluate equations (4.20) and (4.21) and get

$$\overline{P}([3,\infty))[\alpha] = [k(\lambda_1(\alpha)), k(\lambda_2(\alpha))]. \tag{4.22}$$

This fuzzy probability is shown in Figure 4.3. Maple commands for this figure are in Chapter 30.

To finish this section we now compute the fuzzy mean and the fuzzy variance of the fuzzy Poisson probability mass function. Alpha-cuts of the fuzzy mean, from equation (3.51) of Chapter 3, are

$$\overline{\mu}[\alpha] = \{\sum_{x=0}^{\infty} xh(\lambda)|\lambda \in \overline{\lambda}[\alpha]\}, \tag{4.23}$$

which reduces to, since the mean of the crisp Poisson is λ, the expression

$$\overline{\mu}[\alpha] = \{\lambda|\lambda \in \overline{\lambda}[\alpha]\}. \tag{4.24}$$

Hence, $\overline{\mu} = \overline{\lambda}$. So the fuzzy mean is just the fuzzification of the crisp mean.

Let the fuzzy variance be $\overline{\sigma}^2$ and we obtain its α-cuts as

$$\overline{\sigma}^2[\alpha] = \{\sum_{x=0}^{\infty} (x-\mu)^2 h(\lambda)|\lambda \in \overline{\lambda}[\alpha], \mu = \lambda\}, \tag{4.25}$$

which reduces to, since the variance of the crisp Poisson is also λ, the expression

$$\overline{\sigma}^2[\alpha] = \{\lambda|\lambda \in \overline{\lambda}[\alpha]\}. \tag{4.26}$$

It follows that $\overline{\sigma}^2 = \overline{\lambda}$ and the fuzzy variance is the fuzzification of the crisp variance.

4.4 Applications

In this section we look at three applications: (1) using the fuzzy Poisson to approximate values of the fuzzy binomial; (2) using the fuzzy binomial to calculate the fuzzy probabilities of "overbooking"; and (3) then using the fuzzy Poisson to estimate the size of a rapid response team to terrorist attacks.

4.4.1 Fuzzy Poisson Approximating Fuzzy Binomial

Let X be a random variable having the binomial probability mass function $b(n, p)$. From crisp probability theory [1] we know that if n is large and p is small we can use the Poisson to approximate values of the binomial. For non-negative integers a and b, $0 \le a \le b$, let $P([a, b])$ be the probability that $a \le X \le b$. Then using the binomial we have ($q = 1 - p$)

$$P([a, b]) = \sum_{x=a}^{b} \binom{n}{x} p^x q^{n-x}. \tag{4.27}$$

Using the Poisson, with $\lambda = np$, we calculate

$$P([a, b]) \approx \sum_{x=a}^{b} \frac{\lambda^x \exp(-\lambda)}{x!}. \tag{4.28}$$

Now switch to the fuzzy case. Let $\overline{p}$ be small, which means that all $p \in \overline{p}[0]$ are sufficiently small. Let $\overline{P}([a, b])$ be the fuzzy probability that $a \le X \le b$. For .notational simplicity set $\overline{P}_{b\alpha} = \overline{P}([a, b])[\alpha]$ using the fuzzy binomial. Also set $\overline{P}_{p\alpha} = \overline{P}([a, b])[\alpha]$ using the fuzzy Poisson approximation. Then

$$\overline{P}_{b\alpha} = \{\sum_{x=a}^{b} \binom{n}{x} p^x (1 - p)^{n-x} | p \in \overline{p}[\alpha]\}, \tag{4.29}$$

and

$$\overline{P}_{p\alpha} = \{\sum_{x=a}^{b} \frac{\lambda^x \exp(-\lambda)}{x!} | \lambda \in n\overline{p}[\alpha]\}. \tag{4.30}$$

Notice that in equation (4.29) we are using a slightly different model of the fuzzy binomial from equation (4.2) which is similar to, but not exactly equal to, $\overline{q} = 1 - \overline{p}$. We now argue that $\overline{P}_{b\alpha} \approx \overline{P}_{p\alpha}$ for all α. This approximation is to be interpreted as follows: (1) given $z \in \overline{P}_{b\alpha}$, there is a $y \in \overline{P}_{p\alpha}$ so that $z \approx y$; and (2) given $y \in \overline{P}_{p\alpha}$ there is a $z \in \overline{P}_{b\alpha}$ so that $y \approx z$. Also, $z \approx y$ and $y \approx z$ are to be interpreted as in crisp probability theory. To show (1) let $z \in \overline{P}_{b\alpha}$, then $z = \sum_{x=a}^{b} \binom{n}{x} p^x (1-p)^{n-x}$ for some $p \in \overline{p}[\alpha]$. For this same p let $\lambda = np$ and set $y = \sum_{x=a}^{b} \frac{\lambda^x \exp(-\lambda)}{x!}$. Then $z \approx y$. Similarly we show (2).

α	$\overline{P}_{b\alpha}$	$\overline{P}_{p\alpha}$
0	[0.647,0.982]	[0.647,0.981]
0.2	[0.693,0.967]	[0.692,0.966]
0.4	[0.737,0.948]	[0.736,0.946]
0.6	[0.780,0.923]	[0.779,0.921]
0.8	[0.821,0.893]	[0.819,0.891]
1.0	0.859	0.857

Table 4.1: Fuzzy Poisson Approximation to Fuzzy Binomial

Example 4.4.1.1

Let $n = 100$ and $p = 0.02$. Then set $\overline{p} = (0.01/0.02/0.03)$. Now let $a = 0$ and $b = 3$ so $\overline{P}_{b\alpha} = \overline{P}([0,3])[\alpha]$ using the fuzzy binomial and $\overline{P}_{p\alpha} = \overline{P}([0,3])[\alpha]$ using the fuzzy Poisson approximation. We have computed values of $\overline{P}_{b\alpha}$ and $\overline{P}_{p\alpha}$ for $\alpha = 0, 0.2, 0.4, 0.6, 0.8, 1$ and these are shown in Table 4.1. The data in this table shows that, for this example, the fuzzy Poisson is a good approximation to the fuzzy binomial.

To compute $\overline{P}_{b\alpha}$ we simply graphed the function $F(p) = \sum_{x=0}^{3} \binom{n}{x} p^x (1 - p)^{n-x}$ for p in the interval $\overline{p}[\alpha]$, using the software package Maple [2], to pick out the end points of the α-cut. It turns out that $F(p)$ is a decreasing function of p over the interval $\overline{p}[0]$.

Computing $\overline{P}_{p\alpha}$ was easier. Let $G(\lambda) = \sum_{x=0}^{3} \frac{\lambda^x \exp(-\lambda)}{x!}$. We see that $dG/d\lambda < 0$ so if $\overline{\lambda}[\alpha] = [\lambda_1(\alpha), \lambda_2(\alpha)]$, then $\overline{P}_{p\alpha} = [G(\lambda_2(\alpha)), G(\lambda_1(\alpha))]$. Here we use $\overline{\lambda} = n\overline{p} = (1/2/3)$.

4.4.2   Overbooking

Americana Air has the policy of booking as many as 120 persons on an airplane that can seat only 114. Past data implies that approximately only 85% of the booked passengers actually arrive for the flight. We want to find the probability that if Americana Air books 120 persons, not enough seats will be available.

This is a binomial situation with $p \approx 0.85$. Since p has been estimated from past data we use a set of confidence intervals, see Section 3.2, to construct a fuzzy number $\overline{p} = (0.75/0.85/0.95)$ for p producing the fuzzy binomial. Let $\overline{P}_0$ be the fuzzy probability of being overbooked, then its α-cuts are

$$\overline{P}_0[\alpha] = \{ \sum_{x=115}^{120} \binom{120}{x} p^x (1 - p)^{120-x} | p \in \overline{p}[\alpha] \}. \tag{4.31}$$

Again, as in the previous subsection, we are using a slightly different form of the fuzzy binomial than given in Section 4.2. The graph of the fuzzy probability of overbooking is shown in Figure 4.4. Let $F(p) = \sum_{x=115}^{120} \binom{120}{x} p^x (1 -$

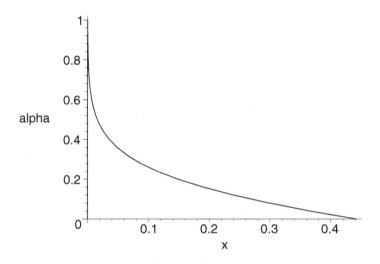

Figure 4.4: Fuzzy Probability of Overbooking

$p)^{120-x}$ for $p \in \overline{p}[0]$. We graphed $F(p)$ using Maple and found that this function is an increasing function of p on the interval $\overline{p}[0]$. This made it easy to evaluate equation (4.31) and obtain the graph in Figure 4.4. Figure 4.4 does not show the left side of the fuzzy number because the left side of the α-cuts involve very small numbers. Selected α-cuts of $\overline{P}_0$ are: (1) $[0.9(10)^{-9}, 0.4415]$ for $\alpha = 0$; (2) $[0.5(10)^{-6}, 0.0160]$ for $\alpha = 0.5$; and (3) $[0.00014, 0.00014]$ for $\alpha = 1$. Maple commands for Figure 4.4 are in Chapter 30.

Notice that the core of $\overline{P}_0$, where the membership is one, is just the crisp probability of overbooking using $p = 0.85$. The spread of the fuzzy number $\overline{P}_0$ shows the uncertainty about the crisp result.

4.4.3 Rapid Response Team

The US government is planning a rapid response team to terrorist attacks within continental US. They need to compute the probability of multiple attacks in a single day to see if they will need one team or multiple teams. It is difficult to do, but they estimate that the mean number of terrorist attacks per day is approximately $\lambda = 0.008$, or about 3 per year starting in 2006. Using the Poisson probability mass function, find the probability that the number of attacks in one day is 0, or 1, or at least 2.

The value of λ was estimated by a group of experts and is very uncertain. Hence we will use a fuzzy number $\overline{\lambda} = (0.005/0.007, 0.009/0.011)$, a trapezoidal fuzzy number, for λ. Let $\overline{P}_m$ be the fuzzy probability of 2 or more attacks per day, which will be used to see if multiple rapid response teams

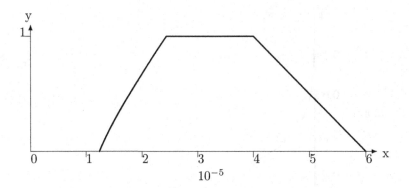

Figure 4.5: Fuzzy Probability of Multiple Attacks

will be needed. Alpha-cuts of this fuzzy probability are

$$\overline{P}_m[\alpha] = \{1 - \sum_{x=0}^{1} \frac{\lambda^x \exp(-\lambda)}{x!} | \lambda \in \overline{\lambda}[\alpha]\}, \tag{4.32}$$

for $0 \le \alpha \le 1$. Let $v(\lambda) = 1 - \sum_{x=0}^{1} \frac{\lambda^x \exp(-\lambda)}{x!}$. We find that $dv/d\lambda > 0$ so if $\overline{\lambda}[\alpha] = [\lambda_1(\alpha), \lambda_2(\alpha)]$, then

$$\overline{P}_m[\alpha] = [v(\lambda_1(\alpha)), v(\lambda_2(\alpha))]. \tag{4.33}$$

The graph of $\overline{P}_m$ is in Figure 4.5. Notice that this fuzzy number is a trapezoidal shaped fuzzy number. Do they need multiple rapid response teams?

4.5   References

1. R.V. Hogg and E.A. Tanis: Probability and Statistical Inference, Sixth Edition, Prentice Hall, Upper Saddle River, N.J., 2001.

2. Maple 9, Waterloo Maple Inc., Waterloo, Canada.

Chapter 5

Continuous Fuzzy Random Variables

5.1 Introduction

We consider the fuzzy uniform in Section 5.2, the fuzzy normal is in Section 5.3, followed by the fuzzy negative exponential in Section 5.4. In each case of a fuzzy density function we first discuss how they are used to compute fuzzy probabilities and then we find their fuzzy mean and their fuzzy variance. We always substitute fuzzy numbers for the parameters in these probability density functions, to produce fuzzy probability density functions. Fuzzy estimators for these parameters will be discussed in: (1) Chapter 11 for the uniform; (2) Chapters 6,7 and 9 for the normal; and (3) Chapter 10 for the exponential.

We will denote the normal probability density as $N(\mu, \sigma^2)$ and the fuzzy normal density as $N(\overline{\mu}, \overline{\sigma}^2)$. The uniform density on interval $[a, b]$ is written $U(a, b)$ and the fuzzy uniform $U(\overline{a}, \overline{b})$ for fuzzy numbers $\overline{a}$ and $\overline{b}$. The negative exponential is $E(\lambda)$ with fuzzy form $E(\overline{\lambda})$.

5.2 Fuzzy Uniform

The uniform density $U(a, b)$, $a < b$, has $y = f(x; a, b) = 1/(b - a)$ for $a \leq x \leq b$ and $f(x; a, b) = 0$ otherwise. Now consider $U(\overline{a}, \overline{b})$ for fuzzy numbers $\overline{a}$ and $\overline{b}$. If $\overline{a}[1] = [a_1, a_2]$ and $\overline{b}[1] = [b_1, b_2]$ we assume that $a \in [a_1, a_2]$, $b \in [b_1, b_2]$ so that $\overline{a}$ ($\overline{b}$) represents the uncertainty in a (b). Now using the fuzzy uniform density we wish to compute the fuzzy probability of obtaining a value in the interval $[c, d]$. Denote this fuzzy probability as $\overline{P}[c, d]$. We can easily generalize to $\overline{P}[E]$ for more general subsets E.

James J. Buckley: *Fuzzy Probability and Statistics*, StudFuzz **196**, 61–74 (2006)
www.springerlink.com

There is uncertainty in the end points of the uniform density but there is no uncertainty in the fact that we have a uniform density. What this means is that given any $s \in \overline{a}[\alpha]$ and $t \in \overline{b}[\alpha]$, $s < t$, we have a $U(s,t)$, or $f(x; s, t) = 1/(t - s)$ on $[s, t]$ and it equals zero otherwise , for all $0 \le \alpha \le 1$. This enables us to find fuzzy probabilities. Let $L(c, d; s, t)$ be the length of the interval $[s, t] \cap [c, d]$. Then

$$\overline{P}[c, d][\alpha] = \{L(c, d; s, t)/(t - s)|s \in \overline{a}[\alpha], t \in \overline{b}[\alpha], s < t\}, \qquad (5.1)$$

for all $\alpha \in [0, 1]$. Equation (5.1) defines the α-cuts and we put these α-cuts together to obtain the fuzzy set $\overline{P}[c, d]$. To find an α-cut of $\overline{P}[c, d]$ we find the probability of getting a value in the interval $[c, d]$ for each uniform density $U(s, t)$ for all $s \in \overline{a}[\alpha]$ and all $t \in \overline{b}[\alpha]$, with $s < t$.

Example 5.2.1

Let $\overline{a} = (0/1/2)$ and $\overline{b} = (3/4/5)$ and $[c, d] = [1, 4]$. Now $\overline{P}[c, d][\alpha] = [p_1(\alpha), p_2(\alpha)]$ an interval whose end points are functions of α. Then $p_1(\alpha)$ is the minimum value of the expression on the right side of equation (5.1) and $p_2(\alpha)$ is the maximum value. That is

$$p_1(\alpha) = min\{L(1, 4; s, t)/(t - s)|s \in \overline{a}[\alpha], t \in \overline{b}[\alpha]\}, \qquad (5.2)$$

and

$$p_2(\alpha) = max\{L(1, 4; s, t)/(t - s)|s \in \overline{a}[\alpha], t \in \overline{b}[\alpha]\}. \qquad (5.3)$$

It is easily seen that $p_2(\alpha) = 1$ all α in this example. To find the minimum we must consider four cases. First $\overline{a}[\alpha] = [\alpha, 2 - \alpha]$ and $\overline{b}[\alpha] = [3 + \alpha, 5 - \alpha]$. Then the cases are: (1) $\alpha \le s \le 1, 3 + \alpha \le t \le 4$; (2) $\alpha \le s \le 1, 4 \le t \le 5 - \alpha$; (3) $1 \le s \le 2 - \alpha, 3 + \alpha \le t \le 4$; and (4) $1 \le s \le 2 - \alpha, 4 \le t \le 5 - \alpha$. Studying all four cases we obtain the minimum equal to $3/(5 - 2\alpha)$. Hence the α-cuts of $\overline{P}[1, 4]$ are $[3/(5 - 2\alpha), 1]$ and the graph of this fuzzy number is in Figure 5.1.

Next we want to find the mean and variance of $U(\overline{a}, \overline{b})$. Let the mean be $\overline{\mu}$ and we find its α-cuts as follows

$$\overline{\mu}[\alpha] = \{\int_s^t (x/(t - s))dx|s \in \overline{a}[\alpha], t \in \overline{b}[\alpha], s < t\}, \qquad (5.4)$$

for all α. But each integral in equation (5.4) equals $(s + t)/2$. Hence, assuming $\overline{a}[0] = [s_1, s_2]$, $\overline{b}[0] = [t_1, t_2]$ and $s_2 < t_1$,

$$\overline{\mu} = (\overline{a} + \overline{b})/2. \qquad (5.5)$$

So, $\overline{\mu}$ is the fuzzification of the crisp mean $(a + b)/2$. If the variance of $U(\overline{a}, \overline{b})$ is $\overline{\sigma}^2$, then its α-cuts are

$$\overline{\sigma}^2[\alpha] = \{\int_s^t [(x - \mu)^2/(t - s)]dx|s \in \overline{a}[\alpha], t \in \overline{b}[\alpha], \mu = (s + t)/2, s < t\}, \quad (5.6)$$

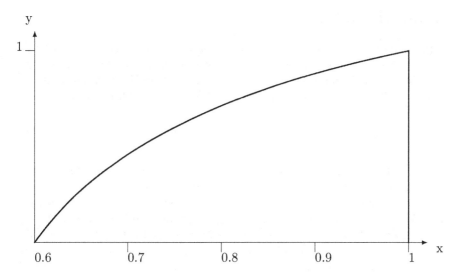

Figure 5.1: Fuzzy Probability in Example 5.2.1

for all α. Each integral in equation (5.6) equals $(t-s)^2/12$. Hence $\overline{\sigma}^2 = (\overline{b}-\overline{a})^2/12$, the fuzzification of the crisp variance.

Next we look at the fuzzy normal probability density.

5.3 Fuzzy Normal

The normal density $N(\mu,\sigma^2)$ has density function $f(x;\mu,\sigma^2)$, $x \in \mathbf{R}$, mean μ and variance σ^2. So consider the fuzzy normal $N(\overline{\mu},\overline{\sigma}^2)$ for fuzzy numbers $\overline{\mu}$ and $\overline{\sigma}^2 > 0$. We wish to compute the fuzzy probability of obtaining a value in the interval $[c,d]$. We write this fuzzy probability as $\overline{P}[c,d]$. We may easily extend our results to $\overline{P}[E]$ for other subsets E of $\mathbf{R}$. For $\alpha \in [0,1]$, $\mu \in \overline{\mu}[\alpha]$ and $\sigma^2 \in \overline{\sigma}^2[\alpha]$ let $z_1 = (c-\mu)/\sigma$ and $z_2 = (d-\mu)/\sigma$. Then

$$\overline{P}[c,d][\alpha] = \{\int_{z_1}^{z_2} f(x;0,1)dx | \mu \in \overline{\mu}[\alpha], \sigma^2 \in \overline{\sigma}^2[\alpha]\}, \qquad (5.7)$$

for $0 \le \alpha \le 1$. The above equation gets the α-cuts of $\overline{P}[c,d]$. Also, in the above equation $f(x;0,1)$ stands for the standard normal density with zero mean and unit variance. Let $\overline{P}[c,d][\alpha] = [p_1(\alpha), p_2(\alpha)]$. Then the minimum (maximum) of the expression on the right side of the above equation is $p_1(\alpha)$ $(p_2(\alpha))$. In general, it could be difficult to find these minimums (maximums) and one might consider using some numerical software (Maple). However, as the following example shows, in some cases we can easily compute these α-cuts.

Example 5.3.1

Suppose $\overline{\mu} = (8/10/12)$, or the mean is approximately 10, and $\overline{\sigma}^2 = (4/5/6)$, or the variance is approximately five. Compute $\overline{P}[10, 15]$. First it is easy to find the $\alpha = 1$ cut and we obtain $\overline{P}[10, 15][1] = 0.4873$. Now we want the $\alpha = 0$ cut. Using the software package Maple [3] we graphed the function

$$g(x, y) = \int_{z_1}^{z_2} f(u; 0, 1)du, \tag{5.8}$$

for $z_1 = (10 - x)/y$, $z_2 = (15 - x)/y$, $8 \le x \le 12$, $4 \le y^2 \le 6$. Notice that the $\alpha = 0$ cut of $(8/10/12)$ is $[8, 12]$, the range for $x = \mu$, and of $(4/5/6)$ is $[4, 6]$ the range for $y^2 = \sigma^2$. The surface clearly shows: (1) a minimum of 0.1584 at $x = 8$ and $y = 2$; and (2) a maximum of 0.7745 at $x = 12$ and $y = 2$. Hence the $\alpha = 0$ cut of this fuzzy probability is $[0.1584, 0.7745]$. But from this graph we may also find other α-cuts. We see from the graph that $g(x, y)$ is an increasing function of: (1) x for y fixed at a value between 2 and $\sqrt{6}$; and (2) y for x fixed at 8. However, $g(x, y)$ is a decreasing function of y for $x = 12$. This means that for any α-cut: (1) we get the max at $y =$ its smallest value and $x =$ at its largest value; and (2) we have the min when $y =$ at is smallest and $x =$ its least value. Some α-cuts of $\overline{P}[10, 15]$ are shown in Table 5.1 and Figure 5.2 displays this fuzzy probability. The graph in Figure 5.2 is only an approximation because we did not force the graph through all the points in Table 5.1.

α	$\overline{P}[10, 15][\alpha]$
0	[0.1584,0.7745]
0.2	[0.2168,0.7340]
0.4	[0.2821,0,6813]
0.6	[0.3512,0.6203]
0.8	[0.4207,0.5545]
1.0	[0.4873,0.4873]

Table 5.1: Alpha-Cuts of the Fuzzy Probability in Example 5.3.1

When [1] was originally published the author's copy of Maple did not contain a nonlinear optimization solver. So we used the graphical method of solution in this example. Now Maple 9 [3] has a non-linear optimization procedure and the Maple commands for this example are in Chapter 30.

We now show that the fuzzy mean of $N(\overline{\mu}, \overline{\sigma}^2)$ is $\overline{\mu}$ and the fuzzy variance is $\overline{\sigma}^2$, respectively, the fuzzification of the crisp mean and variance. Let the fuzzy mean be $\overline{M}$. Then its α-cuts are

$$\overline{M}[\alpha] = \{\int_{-\infty}^{\infty} xf(x; \mu, \sigma^2)dx | \mu \in \overline{\mu}[\alpha], \sigma^2 \in \overline{\sigma}^2[\alpha]\}. \tag{5.9}$$

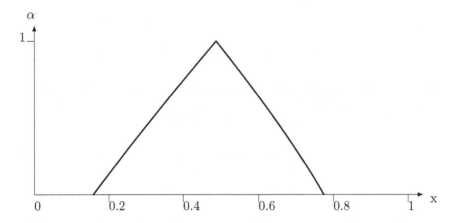

Figure 5.2: Fuzzy Probability in Example 5.3.1

But the integral in the above equation equals μ for any $\mu \in \overline{\mu}[\alpha]$ and any $\sigma^2 \in \overline{\sigma}^2[\alpha]$. Hence $\overline{M} = \overline{\mu}$. Let the fuzzy variance be $\overline{V}$. Then its α-cuts are

$$\overline{V}[\alpha] = \{ \int_{-\infty}^{\infty} (x - \mu)^2 f(x, \mu, \sigma^2) dx | \mu \in \overline{\mu}[\alpha], \sigma^2 \in \overline{\sigma}^2[\alpha], \}. \tag{5.10}$$

We see that the integral in the above equation equals σ^2 for all $\mu \in \overline{\mu}[\alpha]$ and all $\sigma^2 \in \overline{\sigma}^2[\alpha]$. Therefore, $\overline{V} = \overline{\sigma}^2$.

5.4 Fuzzy Negative Exponential

The negative exponential $E(\lambda)$ has density $f(x; \lambda) = \lambda \exp(-\lambda x)$ for $x \geq 0$ and $f(x; \lambda) = 0$ otherwise, where $\lambda > 0$. The mean and variance of $E(\lambda)$ is $1/\lambda$ and $1/\lambda^2$, respectively. Now consider $E(\overline{\lambda})$ for fuzzy number $\overline{\lambda} > 0$. Let us find the fuzzy probability of obtaining a value in the interval $[c, d]$, $c > 0$. Denote this probability as $\overline{P}[c, d]$. One may generalize to $\overline{P}[E]$ for other subsets E of $\mathbf{R}$. We compute

$$\overline{P}[c, d][\alpha] = \{ \int_{c}^{d} \lambda \exp(-\lambda x) dx | \lambda \in \overline{\lambda}[\alpha] \}, \tag{5.11}$$

for all α. Let $\overline{P}[c, d][\alpha] = [p_1(\alpha), p_2(\alpha)]$, then

$$p_1(\alpha) = min\{ \int_{c}^{d} \lambda \exp(-\lambda x) dx | \lambda \in \overline{\lambda}[\alpha] \}, \tag{5.12}$$

and

$$p_2(\alpha) = max\{ \int_{c}^{d} \lambda \exp(-\lambda x) dx | \lambda \in \overline{\lambda}[\alpha] \}, \tag{5.13}$$

for $0 \leq \alpha \leq 1$. Let

$$h(\lambda) = \exp(-c\lambda) - \exp(-d\lambda) = \int_c^d \lambda \exp(-\lambda x) dx, \qquad (5.14)$$

and we see that h: (1) is an increasing function of λ for $0 < \lambda < \lambda^*$; and (2) is a decreasing function of λ for $\lambda^* < \lambda$. We find that $\lambda^* = -[ln(c/d)]/(d - c)$. Assume that $\overline{\lambda} > \lambda^*$. So we can now easily find $\overline{P}[c, d]$. Let $\overline{\lambda}[\alpha] = [\lambda_1(\alpha), \lambda_2(\alpha)]$. Then

$$p_1(\alpha) = h(\lambda_2(\alpha)), \qquad (5.15)$$

and

$$p_2(\alpha) = h(\lambda_1(\alpha)). \qquad (5.16)$$

We give a picture of this fuzzy probability in Figure 5.3 when: (1) $c = 1$ and $d = 4$; and (2) $\overline{\lambda} = (1/3/5)$.

Next we find the fuzzy mean and fuzzy variance of $E(\overline{\lambda})$. If $\overline{\mu}$ denotes the mean, we find its α-cuts as

$$\overline{\mu}[\alpha] = \{ \int_0^\infty x\lambda \exp(-\lambda x) dx | \lambda \in \overline{\lambda}[\alpha] \}, \qquad (5.17)$$

for all α. However, each integral in the above equation equals $1/\lambda$. Hence $\overline{\mu} = 1/\overline{\lambda}$. If $\overline{\sigma}^2$ is the fuzzy variance, then we write down an equation to find its α-cuts we obtain $\overline{\sigma}^2 = 1/\overline{\lambda}^2$. The fuzzy mean (variance) is the fuzzification of the crisp mean (variance).

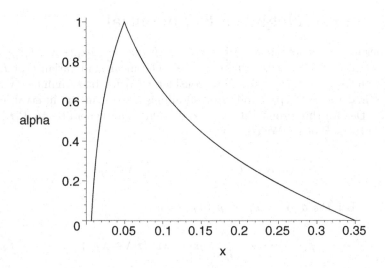

Figure 5.3: Fuzzy Probability for the Fuzzy Exponential

5.5   Applications

In this section we look at some applications of the fuzzy uniform, the fuzzy normal and the fuzzy negative exponential.

5.5.1   Fuzzy Uniform

Customers arrive randomly at a certain shop. Given that one customer arrived during a particular T-minute period, let X be the time within the T minutes that the customer arrived. Assume that the probability density function for X is $U(0,T)$. Find $Prob(4 \leq X \leq 9)$. However, T is not known exactly and is approximately 10, so we will use $\overline{T} = (8/10/12)$ for T. So the probability that $4 \leq X \leq 9$ becomes a fuzzy probability $\overline{P}[4,9]$. Its α-cuts are computed as in equation (5.1). We find that for $0 \leq \alpha \leq 0.5$ that

$$\overline{P}[4,9][\alpha] = \{\frac{min\{t,9\} - 4}{t} | t \in [8 + 2\alpha, 12 - 2\alpha]\}, \qquad (5.18)$$

and for $0.5 \leq \alpha \leq 1$,

$$\overline{P}[4,9][\alpha] = \{\frac{5}{t} | t \in [8 + 2\alpha, 12 - 2\alpha]\}. \qquad (5.19)$$

From this we determine that

$$\overline{P}[4,9][\alpha] = [\frac{5}{12 - 2\alpha}, \frac{5}{9}], \qquad (5.20)$$

for $0 \leq \alpha \leq 0.5$, and

$$\overline{P}[4,9][\alpha] = [\frac{5}{12 - 2\alpha}, \frac{5}{8 + 2\alpha}], \qquad (5.21)$$

for $0.5 \leq \alpha \leq 1$. The graph of this fuzzy probability is in Figure 5.4.

5.5.2   Fuzzy Normal Approximation to Fuzzy Binomial

We first review some basic information about the fuzzy binomial distribution from Chapter 4. Define $X = \{x_1, ..., x_n\}$ and let E be a non-empty, proper, subset of X. We have an experiment where the result is considered a "success" if the outcome x_i is in E. Otherwise, the result is considered a "failure". Let $P(E) = p$ so that $P(E') = q = 1 - p$. $P(E)$ is the probability of success and $P(E')$ is the probability of failure. We assume that $0 < p < 1$.

Suppose we have m independent repetitions of this experiment. If $P(r)$ is the probability of r successes in the m experiments, then

$$P(r) = \binom{m}{r} p^r (1 - p)^{m-r}, \qquad (5.22)$$

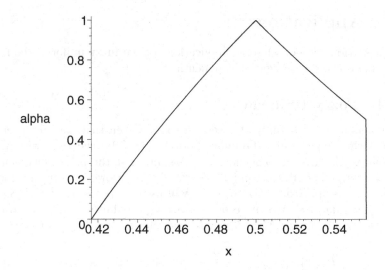

Figure 5.4: Fuzzy Probability $\overline{P}[4,9]$ for the Fuzzy Uniform

for $r = 0, 1, 2, ..., m$, gives the binomial distribution. We write $b(m; p)$ for the crisp binomial and $b(m; \overline{p})$ for the fuzzy binomial. Throughout this section we are using $q = 1 - p$ which is different from the discussion of the fuzzy binomial in Chapter 4.

In these experiments let us assume that $P(E)$ is not known precisely and it needs to be estimated, or obtained from expert opinion. So the p value is uncertain and we substitute $\overline{p}$ for p. Now let $\overline{P}(r)$ be the fuzzy probability of r successes in m independent trials of the experiment. Then

$$\overline{P}(r)[\alpha] = \{ \binom{m}{r} p^r (1-p)^{m-r} | p \in \overline{p}[\alpha] \}, \tag{5.23}$$

for $0 \le \alpha \le 1$. If $\overline{P}(r)[\alpha] = [P_{r1}(\alpha), P_{r2}(\alpha)]$, then

$$P_{r1}(\alpha) = min\{ \binom{m}{r} p^r (1-p)^{m-r} | p \in \overline{p}[\alpha] \}, \tag{5.24}$$

and

$$P_{r2}(\alpha) = max\{ \binom{m}{r} p^r (1-p)^{m-r} | p \in \overline{p}[\alpha] \}. \tag{5.25}$$

Example 5.5.2.1

Let $p = 0.4$ and $m = 3$. Since p is uncertain we use $\overline{p} = (0.3/0.4/0.5)$ for p. Now we will calculate the fuzzy number $\overline{P}(2)$. Equations (5.24) and (5.25) become

$$P_{r1}(\alpha) = min\{3p^2(1-p)|p \in \overline{p}[\alpha]\}, \tag{5.26}$$

and
$$P_{r2}(\alpha) = max\{3p^2(1-p)|p \in \overline{p}[\alpha]\}. \qquad (5.27)$$

Since $d(3p^2(1-p))/dp > 0$ on $\overline{p}[0]$ we obtain

$$\overline{P}(2)[\alpha] = [3(p_1(\alpha))^2(1-p_1(\alpha)), 3(p_2(\alpha))^2(1-p_2(\alpha))], \qquad (5.28)$$

where $\overline{p}[\alpha] = [p_1(\alpha), p_2(\alpha)] = [0.3 + 0.1\alpha, 0.5 - 0.1\alpha]$.

We now need the mean and variance of the fuzzy binomial distribution $b(m; \overline{p})$ which was discussed in Section 4.2. Let $\overline{\mu}$ be the fuzzy mean of the fuzzy binomial and let $\overline{\sigma}^2$ be its fuzzy variance. We showed that, in general, $\overline{\mu} \le m\overline{p}$ and $\overline{\sigma}^2 \le m\overline{p}(1-\overline{p})$. But when we use $q = 1 - p$ we obtain $\overline{\mu} = m\overline{p}$.

Now consider $b(100; \overline{p})$ and we wish to find the fuzzy probability of obtaining from 40 to 60 successes. Denote this fuzzy probability as $\overline{P}[40, 60]$ and direct calculation would be

$$\overline{P}[40, 60][\alpha] = \{\sum_{i=40}^{60} \binom{100}{i} p^i(1-p)^{100-i} | p \in \overline{p}[\alpha]\}, \qquad (5.29)$$

for each α-cut. Now let us try to use the fuzzy normal to approximate this fuzzy probability. Let $f(x; 0, 1)$ be the normal probability density function with zero mean and unit variance. In the following equation $z_1 = (39.5 - \mu)/\sigma$, $z_2 = (60.5 - \mu)/\sigma$, , then

$$\overline{P}[40, 60][\alpha] \approx \{\int_{z_1}^{z_2} f(x; 0, 1)dx | \mu \in \overline{\mu}[\alpha], \sigma^2 \in \overline{\sigma}^2[\alpha]\}, \qquad (5.30)$$

for all α, where $\overline{\mu}$ is the fuzzy mean of the fuzzy binomial and $\overline{\sigma}^2$ is the fuzzy variance of the fuzzy binomial. Let us show that equation (5.30) is correct through the following example.

Example 5.5.2.2

Let $m = 100$, $p \approx 0.6$ so that we use $\overline{p} = (0.5/0.6/0.7)$. For the normal approximation to the binomial to be reasonably accurate one usually assumes that [2] $mp > 5$ and $m(1 - p) > 5$. For the fuzzy normal approximation to the fuzzy binomial to be reasonably good we assume that $m\overline{p} > 5$ and $m(1 - \overline{p}) > 5$, which is true in this example. We now argue that equation (5.30) will give a good approximation to $\overline{P}[40, 60]$. Pick and fix a value of α in $[0, 1)$. Choose $p_0 \in \overline{p}[\alpha]$. Let

$$w = \sum_{i=40}^{60} \binom{100}{i} p_0^i(1-p_0)^{100-i}, \qquad (5.31)$$

with $w \in \overline{P}[40, 60][\alpha]$.

Now we need to compute the fuzzy mean and the fuzzy variance of this fuzzy binomial. We get $\overline{\mu} = 100\overline{p}$. We next compute $\overline{\sigma}^2$ as in Example 4.2.2. We obtain

$$\overline{\sigma}^2[\alpha] = [h(p_2(\alpha), h(p_1(\alpha)], \tag{5.32}$$

where $h(p) = 100p(1-p)$, and $\overline{p}[\alpha] = [p_1(\alpha), p_2(\alpha)] = [0.5 + 0.1\alpha.0.7 - 0.1\alpha]$. The result is $\overline{\sigma}^2[\alpha] = [21 + 4\alpha - \alpha^2, 25 - \alpha^2]$. Then the α-cuts for $\overline{\sigma}$ will be the square root of the α-cuts of $\overline{\sigma}^2$.

Now let $\mu_0 = 100p_0$ in $100\overline{p}[\alpha]$ and let $\sigma_0 \in \overline{\sigma}[\alpha]$ which was computed above. Then

$$w \approx \int_{z_1}^{z_2} f(x; 0, 1)dx, \tag{5.33}$$

where $z_1 = (39.5 - \mu_0)/\sigma_0$, $z_2 = (60.5 - \mu_0)/\sigma_0$.

Now we turn it around and first pick $\mu_0 \in 100\overline{p}[\alpha]$ and $\sigma_0 \in \overline{\sigma}[\alpha]$. But this determines a $p_0 \in \overline{p}[\alpha]$, which then gives a value for w in equation(5.31). The approximation in equation (5.30) now holds.

So we see that under reasonable assumptions the fuzzy normal can approximate the fuzzy binomial. Table 5.2 shows the approximation for $\alpha = 0, 0.2, 0.4, 0.6, 0.8, 1$. Let us explain how we determined the values in Table 5.2. First we graphed the function

$$H(p) = \sum_{x=40}^{60} \binom{100}{x} p^x (1-p)^{100-x}, \tag{5.34}$$

for $p \in [0.5, 0.7]$ and found it is a decreasing function of p on this interval. We then easily found the α-cuts for the fuzzy binomial in Table 5.2. We calculated the α-cuts for the fuzzy normal using Maple [3]. The Maple commands are similar to those in the next example which are given in Chapter 30.

5.5.3 Fuzzy Normal Approximation to Fuzzy Poisson

The fuzzy Poisson was discussed in Section 4.3. Let X be a random variable having a Poisson probability mass function so that, if $P(x)$ is the probability

α	$\overline{P}[40, 60][\alpha]$	Normal Approximation
0	[0.0210,0.9648]	[0.0191,0.9780]
0.2	[0.0558,0.9500]	[0.0539,0.9621]
0.4	[0.1235,0.9025]	[0.1228,0.9139]
0.6	[0.2316,0.8170]	[0.2329,0.8254]
0.8	[0.3759,0.6921]	[0.3786,0.6967]
1.0	[0.5379,0.5379]	[0.5406,0.5406]

Table 5.2: Fuzzy Normal Approximation to Fuzzy Binomial

that $X = x$, we have $P(x) = \lambda^x \exp(-\lambda)/x!$, for $x = 0, 1, 2, 3...$ and $\lambda > 0$. We know, if λ is sufficiently large [2], that we can approximate the crisp Poisson with the crisp normal. Let $\lambda = 20$ and let $P(16, 21]$ be the probability that $16 < X \leq 21$. Then

$$P(16, 21] \approx \int_{z_1}^{z_2} f(x; 0, 1)dx, \qquad (5.35)$$

where $z_1 = (16.5 - \lambda)/\sqrt{\lambda}, z_2 = (21.5 - \lambda)/\sqrt{\lambda}$, and $f(x; 0, 1)$ is the normal probability density function with mean zero and variance one. We used the fact that the mean and variance of the crisp Poisson are both equal to λ to define the z_i. In equation (5.35) the exact value using the Poisson is 0.4226 and the normal approximation gives 0.4144. We now argue that we may use the fuzzy normal to approximate the fuzzy Poisson.

Example 5.5.3.1

Let $\overline{\lambda} = (15/20/25)$ and denote the fuzzy probability that $16 < X \leq 21$ as $\overline{P}(16, 21]$ whose α-cuts are

$$\overline{P}(16, 21][\alpha] = \{ \sum_{x=17}^{21} \lambda^x \exp(-\lambda)/x! | \lambda \in \overline{\lambda}[\alpha] \}, \qquad (5.36)$$

for all α in $[0, 1]$. In the following equation $z_1 = (16.5 - \lambda)/\sqrt{\lambda}$ and $z_2 = (21.5 - \lambda)/\sqrt{\lambda}$, then

$$\overline{P}(16, 21][\alpha] \approx \{ \int_{z_1}^{z_2} f(x; 0, 1)dx | \lambda \in \overline{\lambda}[\alpha] \}, \qquad (5.37)$$

for all α. The argument that this equation is correct is the same as that used in the previous subsection for the fuzzy binomial and the fuzzy normal. Table 5.3 shows the approximation for $\alpha = 0, 0.2, 0.4, 0.6, 0.8, 1$. We used Maple [3] to estimate the α-cuts in Table 5.3. We notice that in this example the approximation is quite good. The Maple commands are in Chapter 30.

α	$\overline{P}(16, 21][\alpha]$	Fuzzy Normal Approximation
0	[0.2096,0.4335]	[0.1974,0.4363]
0.2	[0.2577,0.4335]	[0.2420,0.4363]
0.4	0.3073,0.4335]	[0.2896,0.4363]
0.6	[0.3546,0.4335]	[0.3371,0.4363]
0.8	[0.3948,0.4335]	[0.3804,0.4337]
1	[0.4226,0.4226]	[0.4144,0.4144]

Table 5.3: Fuzzy Normal Approximation to Fuzzy Poisson

5.5.4   Fuzzy Normal

This example has been adapted from an example in [4]. Cockpits in fighter jets were originally designed only for men. However, the US Air Force now recognizes that women also make perfectly good pilots of fighter jets. So various cockpit changes were required to better accommodate the new women pilots. The ejection seat used in the fighter jets was originally designed for men who weighted between 140 and 200 pounds. Based on the data they could get on the pool of possible new women pilots their weight was approximated normally distributed with estimated mean of 143 pounds having an estimated standard deviation of 25 pounds. Any women weighing less than 140 pounds, or more than 200 pounds, would have a greater chance of injury if they had to eject. So the US Air Force wanted to know , given a random sample on n possible women pilots, what is the probability that their mean weight is between 140 and 200 pounds. Answers to such questions are important for the possible redesign of the ejection seats.

The mean of 140 pounds, with standard deviation of 25 pounds, are point estimates and to use just these numbers will not show the uncertainty in these estimates. So we will instead use a set of confidence intervals, as described in Chapters 6,7 and 9, to construct fuzzy numbers $\overline{\mu}$, for the mean, and $\overline{\sigma}$, for the standard deviation. Assume $\overline{\mu} = (140/143/146)$ and $\overline{\sigma} = (23/25/27)$. Suppose y is the mean of the weights of the random sample of $n = 36$ possible women pilots. We now want to calculate the fuzzy probability $\overline{P}[140, 200]$ that $140 \leq y \leq 200$ for y having the fuzzy normal with mean $\overline{\mu}$ and standard deviation $\overline{\sigma}/\sqrt{36}$. We therefore need to calculate the α-cuts

$$\overline{P}[140, 200][\alpha] = \{ \int_{z_1}^{z_2} f(x; 0, 1)dx | \mu \in \overline{\mu}[\alpha], \sigma \in \overline{\sigma}[\alpha] \}, \tag{5.38}$$

all α, where $z_1 = 6(140 - \mu)/\sigma$ and $z_2 = 6(200 - \mu)/\sigma$. The value of equation (5.38) is easily found for $\alpha = 1$ and it is 0.7642 . Also, as in Example 5.3.1 we can get the value when $\alpha = 0$. We used Maple [3] to estimate the α-cuts in Table 5.4. The graph of this fuzzy probability is shown in Figure 5.5. The graph in Figure 5.5 is not completely accurate because we did not force it to go through all the points given in Table 5.4.

α	$\overline{P}[140, 200][\alpha]$
0	[0.5000,0.9412]
0.2	[0.5538,0.9169]
0.4	[0.6083,0,8869]
0.6	[0.6622,0.8511]
0.8	[0.7146,0.8100]
1.0	[0.7642,0.7642]

Table 5.4: Alpha-Cuts of the $\overline{P}[140, 200]$

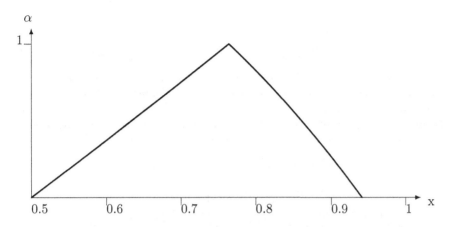

Figure 5.5: Fuzzy Probability in the Ejection Seat Example

5.5.5 Fuzzy Negative Exponential

The crisp negative exponential probability density function is related to the crisp Poisson probability mass function and the same is true in the fuzzy case. A machine has a standby unit available for immediate replacement upon failure. Assume that failures occur for these machines at a rate of λ per hour. Let X be a random variable which counts the number of failures during a period of T hours. Assume that X has a Poisson probability mass function and the probability that $X = x$, denoted by $P_T(x)$, is

$$P_T(x) = (\lambda T)^x \exp(-\lambda T)/x!, \tag{5.39}$$

for $x = 0, 1, 2, 3...$ Now let Y be the random variable whose value is the waiting time to the first failure. It is well known [2] that Y has the exponential probability density function so that

$$Prob[Y > t] = \int_t^\infty \lambda \exp(-\lambda x)dx, \tag{5.40}$$

which is the probability that the first failure occurs after t hours.

Now switch to the fuzzy Poisson with $\overline{\lambda} = (0.07/0.1/0.13)$ for λ and denote the fuzzy probability that the first failure occurs after 10 hours as $\overline{P}[10, \infty]$. Then its α-cuts are

$$\overline{P}[10, \infty][\alpha] = \{\int_{10}^\infty \lambda \exp(-\lambda x)dx | \lambda \in \overline{\lambda}[\alpha]\}, \tag{5.41}$$

for all α. These α-cuts are easy to find because the integral in the above equation is simply $\exp(-10\lambda)$ for $\lambda \in \overline{\lambda}[\alpha]$. So, if $\overline{\lambda}[\alpha] = [\lambda_1(\alpha), \lambda_2(\alpha)]$, then

$$\overline{P}[10, \infty][\alpha] = [\exp(-10\lambda_2(\alpha)), \exp(-10\lambda_1(\alpha))], \tag{5.42}$$

which equals

$$\overline{P}[10, \infty][\alpha] = [\exp(-1.3 + 0.3\alpha), \exp(-0.7 - 0.3\alpha)]. \qquad (5.43)$$

To summarize, if we substitute $\overline{\lambda}$ for λ in Equation (5.39) the fuzzy Poisson can be used to find the fuzzy probability of x failures in time interval T and the fuzzy negative exponential gives the fuzzy times between successive failures. An important property of the crisp exponential is its "forgetfulness". The probability statement of this property is

$$Prob[Y > t_1 + t_2 | Y > t_2] = Prob[Y > t_1]. \qquad (5.44)$$

The time interval remaining until the next failure is independent of the time interval that has elapsed since the last failure. We now show this is also true for the fuzzy exponential using our definition of fuzzy conditional probability in Section 3.6. Using the fuzzy negative exponential α-cuts of the fuzzy conditional probability, relating to the left side of equation (5.44), is

$$\overline{P}[Y > t_1 + t_2 | Y > t_2][\alpha] = \{ \frac{\int_{t_1+t_2}^{\infty} \lambda \exp(-\lambda x) dx}{\int_{t_2}^{\infty} \lambda \exp(-\lambda x) dx} | \lambda \in \overline{\lambda}[\alpha] \}, \qquad (5.45)$$

for $\alpha \in [0, 1]$. Now the quotient of the integrals in equation (5.45) equals, after evaluation, $\exp(-t_1 \lambda)$, so

$$\overline{P}[Y > t_1 + t_2 | Y > t_2][\alpha] = \{ \int_{t_1}^{\infty} \lambda \exp(-\lambda x) dx | \lambda \in \overline{\lambda}[\alpha] \}, \qquad (5.46)$$

which equals $\overline{P}[Y > t_1][\alpha]$. Hence, equation (5.44) also holds for the fuzzy negative exponential and it has the "forgetfulness" property.

5.6   References

1. J.J. Buckley: Fuzzy Statistics, Springer, Heidelberg, Germany 2004.

2. R.V. Hoog and E.A. Tanis: Probability and Statistical Inference, Sixth Edition, Prentice Hall, Upper Saddle River, N.J., 2001.

3. Maple 9, Waterloo Maple Inc. Waterloo, Canada.

4. M.F. Triola: Elementary Statistics Using Excel, Second Edition, Addison-Wesley, N.Y., 2004.

Chapter 6

Estimate μ, Variance Known

6.1 Introduction

This starts a series of chapters, Chapters 6-11, on elementary fuzzy estimation. In this chapter we first present some general information on fuzzy estimation and then concentrate on the mean of a normal probability distribution assuming the variance is known. The rest of the chapters on elementary fuzzy estimation can be read independently. More fuzzy estimation is in Chapters 12-14,19,20,23 and 28.

6.2 Fuzzy Estimation

We have been using (Chapters 4 and 5), and will continue to use, fuzzy numbers for estimators of parameters in probability density functions (probability mass functions in the discrete case) and in this section we show how we obtain these fuzzy numbers from a set of confidence intervals. The discussion here is similar to, but more general than, that presented in Section 3.2. Let X be a random variable with probability density function (or probability mass function) $f(x; \theta)$ for single parameter θ. Assume that θ is unknown and it must be estimated from a random sample $X_1, ..., X_n$. Let $Y = u(X_1, ..., X_n)$ be a statistic used to estimate θ. Given the values of these random variables $X_i = x_i$, $1 \le i \le n$, we obtain a point estimate $\theta^* = y = u(x_1, ..., x_n)$ for θ. We would never expect this point estimate to exactly equal θ so we often also compute a $(1 - \beta)100\%$ confidence interval for θ. We are using β here since α, usually employed for confidence intervals, is reserved for α-cuts of fuzzy numbers. In this confidence interval one usually sets β equal to 0.10, 0.05 or 0.01.

James J. Buckley: *Fuzzy Probability and Statistics*, StudFuzz **196**, 75–79 (2006)
www.springerlink.com   © Springer-Verlag Berlin Heidelberg 2006

We propose to find the $(1-\beta)100\%$ confidence interval for all $0.01 \le \beta < 1$. Starting at 0.01 is arbitrary and you could begin at 0.10 or 0.05 or 0.005, etc. Denote these confidence intervals as

$$[\theta_1(\beta), \theta_2(\beta)], \tag{6.1}$$

for $0.01 \le \beta < 1$. Add to this the interval $[\theta^*, \theta^*]$ for the 0% confidence interval for θ. Then we have $(1-\beta)100\%$ confidence intervals for θ for $0.01 \le \beta \le 1$.

Now place these confidence intervals, one on top of the other, to produce a triangular shaped fuzzy number $\overline{\theta}$ whose α-cuts are the confidence intervals. We have

$$\overline{\theta}[\alpha] = [\theta_1(\alpha), \theta_2(\alpha)], \tag{6.2}$$

for $0.01 \le \alpha \le 1$. All that is needed is to finish the "bottom" of $\overline{\theta}$ to make it a complete fuzzy number. We will simply drop the graph of $\overline{\theta}$ straight down to complete its α-cuts so

$$\overline{\theta}[\alpha] = [\theta_1(0.01), \theta_2(0.01)], \tag{6.3}$$

for $0 \le \alpha < 0.01$. In this way we are using more information in $\overline{\theta}$ than just a point estimate, or just a single interval estimate.

6.3   Fuzzy Estimator of μ

Consider X a random variable with probability density function $N(\mu, \sigma^2)$, which is the normal probability density with unknown mean μ and known variance σ^2. To estimate μ we obtain a random sample $X_1, ..., X_n$ from $N(\mu, \sigma^2)$. Suppose the mean of this random sample turns out to be $\overline{x}$, which is a crisp number, not a fuzzy number. We know that $\overline{x}$ is $N(\mu, \sigma^2/n)$ (Section 7.2 in [1]). So $(\overline{x} - \mu)/(\sigma/\sqrt{n})$ is $N(0,1)$. Therefore

$$P(-z_{\beta/2} \le \frac{\overline{x} - \mu}{\sigma/\sqrt{n}} \le z_{\beta/2}) = 1 - \beta, \tag{6.4}$$

where $z_{\beta/2}$ is the z value so that the probability of a $N(0,1)$ random variable exceeding it is $\beta/2$. Now solve the inequality for μ producing

$$P(\overline{x} - z_{\beta/2}\sigma/\sqrt{n} \le \mu \le \overline{x} + z_{\beta/2}\sigma/\sqrt{n}) = 1 - \beta. \tag{6.5}$$

This leads directly to the $(1-\beta)100\%$ confidence interval for μ

$$[\theta_1(\beta), \theta_2(\beta)] = [\overline{x} - z_{\beta/2}\sigma/\sqrt{n}, \overline{x} + z_{\beta/2}\sigma/\sqrt{n}], \tag{6.6}$$

where $z_{\beta/2}$ is defined as

$$\int_{-\infty}^{z_{\beta/2}} N(0,1)dx = 1 - \beta/2, \tag{6.7}$$

and $N(0, 1)$ denotes the normal density with mean zero and unit variance. Put these confidence intervals together as discussed above and we obtain $\overline{\mu}$ our fuzzy estimator of μ.

The following examples show that the fuzzy estimator of the mean of the normal probability density will be a triangular shaped fuzzy number.

Example 6.3.1

Consider X a random variable with probability density function $N(\mu, 100)$, which is the normal probability density with unknown mean μ and known variance $\sigma^2 = 100$. To estimate μ we obtain a random sample $X_1, ..., X_n$ from $N(\mu, 100)$. Suppose the mean of this random sample turns out to be 28.6. Then a $(1 - \beta)100\%$ confidence interval for μ is

$$[\theta_1(\beta), \theta_2(\beta)] = [28.6 - z_{\beta/2}10/\sqrt{n}, 28.6 + z_{\beta/2}10/\sqrt{n}]. \tag{6.8}$$

To obtain a graph of fuzzy μ, or $\overline{\mu}$, let $n = 64$ and first assume that $0.01 \leq \beta \leq 1$. We evaluated equation (6.8) using Maple [2] and then the final graph of $\overline{\mu}$ is shown in Figure 6.1, without dropping the graph straight down to the x-axis at the end points.

Let us go through more detail on how Maple creates the graph of the fuzzy estimator in Figure 6.1. Some information was given in Section 1.4 of Chapter 1 and the Maple commands for selected figures are in Chapter 30. In equation (6.8) the left (right) end point of the interval describes the left (right) side of $\overline{\mu}$. Let the horizontal axis be called the x-axis and the vertical

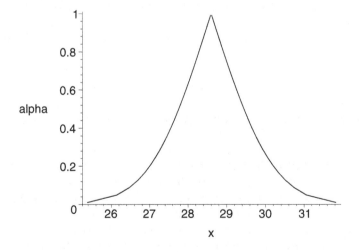

Figure 6.1: Fuzzy Estimator $\overline{\mu}$ in Example 6.3.1, $0.01 \leq \beta \leq 1$

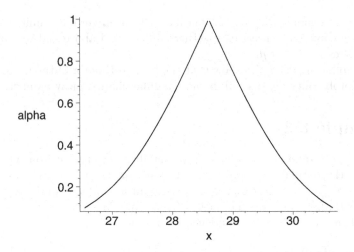

Figure 6.2: Fuzzy Estimator $\overline{\mu}$ in Example 6.3.1, $0.10 \leq \beta \leq 1$

axis the y-axis. We now substitute y for β in equation (6.8). then

$$x = 28.6 - (1.25)z_{y/2}, \tag{6.9}$$

gives the left side of $\overline{\mu}$ and

$$x = 28.6 + (1.25)z_{y/2}, \tag{6.10}$$

is the right side of the fuzzy estimator. We have used $n = 64$ in equation (6.8). But equations (6.9) and (6.10) are "backwards" in that they give x a function of y. But using the "implicitplot" command in Maple equations (6.9) and (6.10) can be graphed and the result for $0.01 \leq y \leq 1$ is Figure 6.1.

We next evaluated equation (6.8) for $0.10 \leq \beta \leq 1$ and then the graph of $\overline{\mu}$ is shown in Figure 6.2, again without dropping the graph straight down to the x-axis at the end points. The Maple commands for Figure 6.2 are in Chapter 30. Finally, we computed equation (6.8) for $0.001 \leq \beta \leq 1$ and the graph of $\overline{\mu}$ is displayed in Figure 6.3 without dropping the graph straight down to the x-axis at the end points.

The graph in Figure 6.2 is a little misleading because the vertical axis does not start at zero. It begins at 0.08. To complete the pictures we draw short vertical line segments, from the horizontal axis up to the graph, at the end points of the base of the fuzzy number $\overline{\mu}$. The base ($\overline{\mu}[0]$) in Figure 6.1 (6.2, 6.3) is a 99% (90%, 99.9%) confidence interval for μ.

In future chapters we usually do not explicitly mention the short vertical line segments at the two ends of the graph connecting the graph and the horizontal axis. Also, in when we used a fuzzy estimator $\overline{\mu}$ in the fuzzy normal (Section 5.3) for simplicity we used a triangular fuzzy number.

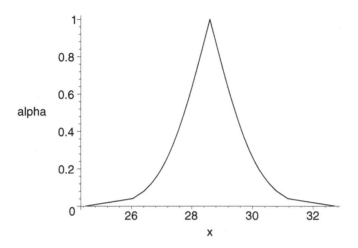

Figure 6.3: Fuzzy Estimator $\overline{\mu}$ in Example 6.3.1, $0.001 \leq \beta \leq 1$

6.4 References

1. R.V. Hogg and E.A. Tanis: Probability and Statistical Inference, Sixth Edition, Prentice Hall, Upper Saddle River, N.J., 2001.

2. Maple 9, Waterloo Maple Inc., Waterloo, Canada.

Chapter 7

Estimate μ, Variance Unknown

7.1 Fuzzy Estimator of μ

Consider X a random variable with probability density function $N(\mu, \sigma^2)$, which is the normal probability density with unknown mean μ and unknown variance σ^2. To estimate μ we obtain a random sample $X_1, ..., X_n$ from $N(\mu, \sigma^2)$. Suppose the mean of this random sample turns out to be $\bar{x}$, which is a crisp number, not a fuzzy number. Also, let s^2 be the sample variance. Our point estimator of μ is $\bar{x}$. If the values of the random sample are $x_1, ..., x_n$ then the expression we will use for s^2 in this book is

$$s^2 = \sum_{i=1}^{n} (x_i - \bar{x})^2 / (n-1). \tag{7.1}$$

We will use this form of s^2, with denominator $(n-1)$, so that it is an unbiased estimator of σ^2.

It is known that $(\bar{x} - \mu)/(s/\sqrt{n})$ has a (Student's) t distribution with $n-1$ degrees of freedom (Section 7.2 of [1]). It follows that

$$P(-t_{\beta/2} \leq \frac{\bar{x} - \mu}{s/\sqrt{n}} \leq t_{\beta/2}) = 1 - \beta, \tag{7.2}$$

where $t_{\beta/2}$ is defined from the (Student's) t distribution, with $n-1$ degrees of freedom, so that the probability of exceeding it is $\beta/2$. Now solve the inequality for μ giving

$$P(\bar{x} - t_{\beta/2}s/\sqrt{n} \leq \mu \leq \bar{x} + t_{\beta/2}s/\sqrt{n}) = 1 - \beta. \tag{7.3}$$

For this we immediately obtain the $(1-\beta)100\%$ confidence interval for μ

$$[\bar{x} - t_{\beta/2}s/\sqrt{n}, \bar{x} + t_{\beta/2}s/\sqrt{n}]. \tag{7.4}$$

James J. Buckley: *Fuzzy Probability and Statistics*, StudFuzz **196**, 81–83 (2006)
www.springerlink.com © Springer-Verlag Berlin Heidelberg 2006

Put these confidence intervals together, as discussed in Chapter 6, and we obtain $\overline{\mu}$ our fuzzy number estimator of μ.

Example 7.1.1

Consider X a random variable with probability density function $N(\mu, \sigma^2)$, which is the normal probability density with unknown mean μ and unknown variance σ^2. To estimate μ we obtain a random sample $X_1, ..., X_n$ from $N(\mu, \sigma^2)$. Suppose the mean of this random sample of size 25 turns out to be 28.6 and $s^2 = 3.42$. Then a $(1 - \beta)100\%$ confidence interval for μ is

$$[28.6 - t_{\beta/2}\sqrt{3.42/25}, 28.6 + t_{\beta/2}\sqrt{3.42/25}]. \qquad (7.5)$$

To obtain a graph of fuzzy μ, or $\overline{\mu}$, first assume that $0.01 \leq \beta \leq 1$. We evaluated equation (7.5) using Maple [2] and then the graph of $\overline{\mu}$ is shown in Figure 7.1, without dropping the graph straight down to the x-axis at the end points. The Maple commands for Figure 7.1 are in Chapter 30.

We next evaluated equation (7.5) for $0.10 \leq \beta \leq 1$ and then the graph of $\overline{\mu}$ is shown in Figure 7.2, again without dropping the graph straight down to the x-axis at the end points. Finally, we computed equation (7.5) for $0.001 \leq \beta \leq 1$ and the graph of $\overline{\mu}$ is displayed in Figure 7.3 without dropping the graph straight down to the x-axis at the end points.

The graph in Figure 7.2 is a little misleading because the vertical axis does not start at zero. It begins at 0.08. To complete the pictures we draw short vertical line segments, from the horizontal axis up to the graph, at the end points of the base of the fuzzy number $\overline{\mu}$. The base ($\overline{\mu}[0]$) in Figure 7.1 (7.2, 7.3) is a 99% (90%, 99.9%) confidence interval for μ.

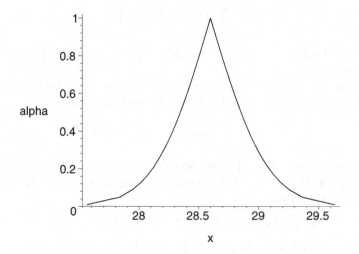

Figure 7.1: Fuzzy Estimator $\overline{\mu}$ in Example 7.1.1, $0.01 \leq \beta \leq 1$

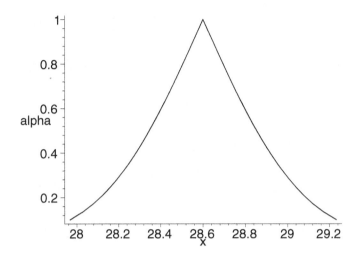

Figure 7.2: Fuzzy Estimator $\overline{\mu}$ in Example 7.1.1, $0.10 \leq \beta \leq 1$

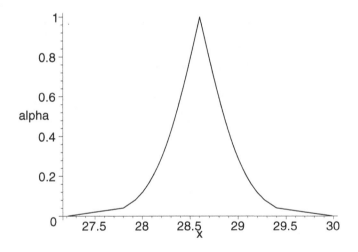

Figure 7.3: Fuzzy Estimator $\overline{\mu}$ in Example 7.1.1, $0.001 \leq \beta \leq 1$

7.2   References

1. R.V. Hogg and E.A. Tanis: Probability and Statistical Inference, Sixth Edition, Prentice Hall, Upper Saddle River, N.J., 2001.

2. Maple 9, Waterloo Maple Inc., Waterloo, Canada.

Chapter 8

Estimate p, Binomial Population

8.1 Fuzzy Estimator of p

This has been previously discussed in Section 3.2, but for completeness it is repeated here with the other fuzzy estimators. We have an experiment in mind in which we are interested in only two possible outcomes labeled "success" and "failure". Let p be the probability of a success so that $q = 1-p$ will be the probability of a failure. We want to estimate the value of p. We therefore gather a random sample which here is running the experiment n independent times and counting the number of times we had a success. Let x be the number of times we observed a success in n independent repetitions of this experiment. Then our point estimate of p is $\widehat{p} = x/n$.

We know that (Section 7.5 in [1]) that $(\widehat{p} - p)/\sqrt{p(1 - p)/n}$ is approximately $N(0,1)$ if n is sufficiently large. Throughout this book we will always assume that the sample size is large enough for the normal approximation to the binomial. Then

$$P(z_{\beta/2} \leq \frac{\widehat{p} - p}{\sqrt{p(1 - p)/n}} \leq z_{\beta/2}) \approx 1 - \beta, \tag{8.1}$$

where $z_{\beta/2}$ was defined in equation (6.7) in Chapter 6. Solving the inequality for the p in the numerator we have

$$P(\widehat{p} - z_{\beta/2}\sqrt{p(1 - p)/n} \leq p \leq \widehat{p} + z_{\beta/2}\sqrt{p(1 - p)/n}) \approx 1 - \beta. \tag{8.2}$$

This leads to the $(1 - \beta)100\%$ approximate confidence interval for p

$$[\widehat{p} - z_{\beta/2}\sqrt{p(1 - p)/n}, \widehat{p} + z_{\beta/2}\sqrt{p(1 - p)/n}]. \tag{8.3}$$

However, we have no value for p to use in this confidence interval. So, still assuming that n is sufficiently large, we substitute $\widehat{p}$ for p in equation (8.3),

James J. Buckley: *Fuzzy Probability and Statistics*, StudFuzz **196**, 85–87 (2006)
www.springerlink.com   © Springer-Verlag Berlin Heidelberg 2006

using $\widehat{q} = 1 - \widehat{p}$, and we get the final $(1 - \beta)100\%$ approximate confidence interval

$$[\widehat{p} - z_{\beta/2}\sqrt{\widehat{pq}/n}, \widehat{p} + z_{\beta/2}\sqrt{\widehat{pq}/n}]. \tag{8.4}$$

Put these confidence intervals together, as discussed in Chapter 6, and we get $\overline{p}$ our triangular shaped fuzzy number estimator of p.

Example 8.1.1

Assume that $n = 350$, $x = 180$ so that $\widehat{p} = 0.5143$. The confidence intervals become

$$[0.5143 - 0.0267z_{\beta/2}, 0.5143 + 0.0267z_{\beta/2}], \tag{8.5}$$

for $0.01(0.10, 0.001) \leq \beta \leq 1$.

To obtain a graph of fuzzy p, or $\overline{p}$, first assume that $0.01 \leq \beta \leq 1$. We evaluated equation (8.5) using Maple [2] and then the graph of $\overline{p}$ is shown in Figure 8.1, without dropping the graph straight down to the x-axis at the end points. The Maple commands for Figure 8.1 are in Chapter 30.

We next evaluated equation (8.5) for $0.10 \leq \beta \leq 1$ and then the graph of $\overline{p}$ is shown in Figure 8.2, again without dropping the graph straight down to the x-axis at the end points. Finally, we computed equation (8.5) for $0.001 \leq \beta \leq 1$ and the graph of is displayed in Figure 8.3 without dropping the graph straight down to the x-axis at the end points.

The graph in Figure 8.2 is a little misleading because the vertical axis does not start at zero. It begins at 0.08. To complete the pictures we draw short vertical line segments, from the horizontal axis up to the graph, at the

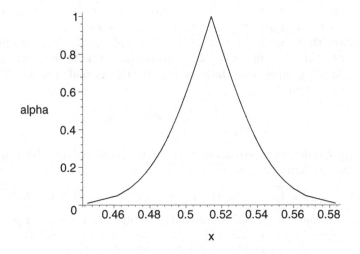

Figure 8.1: Fuzzy Estimator $\overline{p}$ in Example 8.1.1, $0.01 \leq \beta \leq 1$

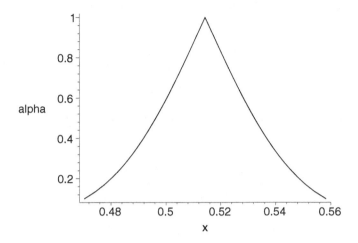

Figure 8.2: Fuzzy Estimator $\bar{p}$ in Example 8.1.1, $0.10 \leq \beta \leq 1$

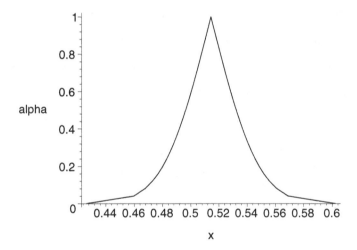

Figure 8.3: Fuzzy Estimator $\bar{p}$ in Example 8.1.1, $0.001 \leq \beta \leq 1$

end points of the base of the fuzzy number $\bar{\mu}$. The base ($\bar{\mu}[0]$) in Figure 8.1 (8.2, 8.3) is a 99% (90%, 99.9%) confidence interval for p.

8.2 References

1. R.V. Hogg and E.A. Tanis: Probability and Statistical Inference, Sixth Edition, Prentice Hall, Upper Saddle River, N.J., 2001.

2. Maple 9, Waterloo Maple Inc., Waterloo, Canada.

Chapter 9

Estimate σ^2 from a Normal Population

9.1 Introduction

We first construct a fuzzy estimator for σ^2 using the usual confidence intervals for the variance from a normal distribution and we show this fuzzy estimator is biased. Then in Section 9.3 we construct an unbiased fuzzy estimator for the variance.

9.2 Biased Fuzzy Estimator

Consider X a random variable with probability density function $N(\mu, \sigma^2)$, which is the normal probability density with unknown mean μ and unknown variance σ^2. To estimate σ^2 we obtain a random sample $X_1, ..., X_n$ from $N(\mu, \sigma^2)$. Our point estimator for the variance will be s^2. If the values of the random sample are $x_1, ..., x_n$ then the expression we will use for s^2 in this book is

$$s^2 = \sum_{i=1}^{n}(x_i - \overline{x})^2/(n-1). \tag{9.1}$$

We will use this form of s^2, with denominator $(n-1)$, so that it is an unbiased estimator of σ^2.

We know that (Section 7.4 in [1]) $(n-1)s^2/\sigma^2$ has a chi-square distribution with $n-1$ degrees of freedom. Then

$$P(\chi^2_{L,\beta/2} \leq (n-1)s^2/\sigma^2 \leq \chi^2_{R,\beta/2}) = 1 - \beta, \tag{9.2}$$

where $\chi^2_{R,\beta/2}$ ($\chi^2_{L,\beta/2}$) is the point on the right (left) side of the χ^2 density where the probability of exceeding (being less than) it is $\beta/2$. The χ^2

James J. Buckley: *Fuzzy Probability and Statistics*, StudFuzz **196**, 89–94 (2006)
www.springerlink.com

distribution has $n-1$ degrees of freedom. Solve the inequality for σ^2 and we see that

$$P(\frac{(n-1)s^2}{\chi^2_{R,\beta/2}} \le \sigma^2 \le \frac{(n-1)s^2}{\chi^2_{L,\beta/2}}) = 1 - \beta. \tag{9.3}$$

From this we obtain the usual $(1-\beta)100\%$ confidence intervals for σ^2

$$[(n-1)s^2/\chi^2_{R,\beta/2}, (n-1)s^2/\chi^2_{L,\beta/2}]. \tag{9.4}$$

Put these confidence intervals together, as discussed in Chapter 6, and we obtain $\overline{\sigma}^2$ our fuzzy number estimator of σ^2.

We now show that this fuzzy estimator is biased because the vertex of the triangular shaped fuzzy number $\overline{\sigma}^2$, where the membership value equals one, is not at s^2. We say a fuzzy estimator is biased when its vertex is not at the point estimator. We obtain the vertex of $\overline{\sigma}^2$ when $\beta = 1.0$. Let

$$factor = \frac{n-1}{\chi^2_{R,0.50}} = \frac{n-1}{\chi^2_{L,0.50}}, \tag{9.5}$$

after we substitute $\beta = 1$. Then the 0% confidence interval for the variance is

$$[(factor)(s^2), (factor)(s^2)] = (factor)(s^2). \tag{9.6}$$

Since $factor \ne 1$ the fuzzy number $\overline{\sigma}^2$ is not centered at s^2. Table 9.1 shows some values of $factor$ for various choices for n. We see that $factor \to 1$ as $n \to \infty$ but $factor$ is substantially larger than one for small values on n. This fuzzy estimator is biased and we will construct an unbiased (vertex at s^2) in the next section.

n	$factor$
10	1.0788
20	1.0361
50	1.0138
100	1.0068
500	1.0013
1000	1.0007

Table 9.1: Values of $factor$ for Various Values of n

9.3   Unbiased Fuzzy Estimator

In deriving the usual confidence interval for the variance we start with recognizing that $(n-1)s^2/\sigma^2$ has a χ^2 distribution with $n-1$ degrees of freedom. Then for a $(1-\beta)100\%$ confidence interval we may find a and b so that

$$P(a \le \frac{(n-1)s^2}{\sigma^2} \le b) = 1 - \beta. \tag{9.7}$$

The usual confidence interval has a and b so that the probabilities in the "two tails" are equal. That is, $a = \chi^2_{L,\beta/2}$ ($b = \chi^2_{R,\beta/2}$) so that the probability of being less (greater) than a (b) is $\beta/2$. But we do not have to pick the a and b this way ([1], p. 378). We will change the way we pick the a and b so that the fuzzy estimator is unbiased.

Assume that $0.01 \leq \beta \leq 1$. Now this interval for β is fixed and also n and s^2 are fixed. Define

$$L(\lambda) = [1 - \lambda]\chi^2_{R,0.005} + \lambda(n - 1), \tag{9.8}$$

and

$$R(\lambda) = [1 - \lambda]\chi^2_{L,0.005} + \lambda(n - 1). \tag{9.9}$$

Then a confidence interval for the variance is

$$[\frac{(n-1)s^2}{L(\lambda)}, \frac{(n-1)s^2}{R(\lambda)}], \tag{9.10}$$

for $0 \leq \lambda \leq 1$. We start with a 99% confidence interval when $\lambda = 0$ and end up with a 0% confidence interval for $\lambda = 1$. Notice that now the 0% confidence interval is $[s^2, s^2] = s^2$ and it is unbiased. As usual, we place these confidence intervals one on top of another to obtain our (unbiased) fuzzy estimator $\overline{\sigma}^2$ for the variance. Our confidence interval for σ, the population standard deviation, is

$$[\sqrt{(n-1)/L(\lambda)}s, \sqrt{(n-1)/R(\lambda)}s]. \tag{9.11}$$

Let us compare the methods in this section to those in Section 9.2. Let χ^2 be the chi-square probability density with $n - 1$ degrees of freedom. The mean of χ^2 is $n - 1$ and the median is the point md where $P(X \leq md) = P(X \geq md) = 0.5$. We assume β is in the interval $[0.01, 1]$. In Section 9.2 as β continuously increases from 0.01 to 1, $\chi^2_{L,\beta/2}$ ($\chi^2_{R,\beta/2}$) starts at $\chi^2_{L,0.005}$ ($\chi^2_{R,0.005}$) and increases (decreases) to $\chi^2_{L,0.5}$ ($\chi^2_{R,0.5}$) which equals to the median. Recall that $\chi^2_{L,\beta/2}$ ($\chi^2_{R,\beta/2}$) is the point on the χ^2 density where the probability of being less (greater) that it equals $\beta/2$. From Table 9.1 we see that the median is always less than $n - 1$. This produces the bias in the fuzzy estimator in that section. In this section as λ continuously increases from zero to one $L(\lambda)$ ($R(\lambda)$) decreases (increases) from $\chi^2_{R,0.005}$ ($\chi^2_{L,0.005}$) to $n - 1$. At $\lambda = 1$ we get $L(1) = R(1) = n - 1$ and the vertex (membership value one) is at s^2 and it is now unbiased.

We will use this fuzzy estimator $\overline{\sigma}^2$ constructed in this section for σ^2 in the rest of this book. Given a value of $\lambda = \lambda^* \in [0, 1]$ one may wonder what is the corresponding value of β for the confidence interval. We now show how to get the β. Let $L^* = L(\lambda^*)$ and $R^* = R(\lambda^*)$. Define

$$l = \int_0^{R^*} \chi^2 dx, \tag{9.12}$$

and
$$r = \int_{L^*}^{\infty} \chi^2 dx, \tag{9.13}$$
and then $\beta = l + r$. Now l (r) need not equal $\beta/2$. Both of these integrals above are easily evaluated using Maple [2]. The chi-square density inside these integrals has $n - 1$ degrees of freedom.

Example 9.3.1

Consider X a random variable with probability density function $N(\mu, \sigma^2)$, which is the normal probability density with mean μ and unknown variance σ^2. To estimate σ^2 we obtain a random sample $X_1, ..., X_n$ from $N(\mu, \sigma^2)$. Suppose $n = 25$ and we calculate $s^2 = 3.42$. Then a confidence interval for σ^2 is
$$[\frac{82.08}{L(\lambda)}, \frac{82.08}{R(\lambda)}]. \tag{9.14}$$
To obtain a graph of fuzzy σ^2, or $\overline{\sigma}^2$, first assume that $0.01 \le \beta \le 1$. We evaluated equation (9.14) using Maple [2] and then the graph of $\overline{\sigma}^2$ is shown in Figure 9.1, without dropping the graph straight down to the x-axis at the end points. The Maple commands for Figure 9.1 are in Chapter 30.

We next evaluated equation (9.14) for $0.10 \le \beta \le 1$ and then the graph of $\overline{\sigma}^2$ is shown in Figure 9.2, again without dropping the graph straight down to the x-axis at the end points. Finally, we computed equation (9.14) for $0.001 \le \beta \le 1$ and the graph of $\overline{\sigma}^2$ is displayed in Figure 9.3 without dropping the graph straight down to the x-axis at the end points.

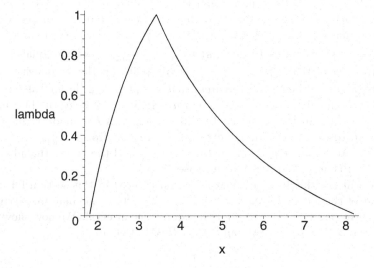

Figure 9.1: Fuzzy Estimator $\overline{\sigma}^2$ in Example 9.3.1, $0.01 \le \beta \le 1$

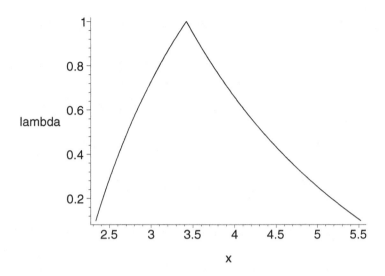

Figure 9.2: Fuzzy Estimator $\overline{\sigma}^2$ in Example 9.3.1, $0.10 \leq \beta \leq 1$

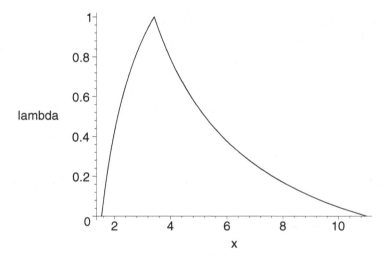

Figure 9.3: Fuzzy Estimator $\overline{\sigma}^2$ in Example 9.3.1, $0.001 \leq \beta \leq 1$

The graph in Figure 9.2 is a little misleading because the vertical axis does not start at zero. It begins at 0.08. To complete the pictures we draw short vertical line segments, from the horizontal axis up to the graph, at the end points of the base of the fuzzy number $\overline{\sigma}^2$. The base ($\overline{\sigma}^2[0]$) in Figure 9.1 (9.2, 9.3) is a 99% (90%, 99.9%) confidence interval for σ^2.

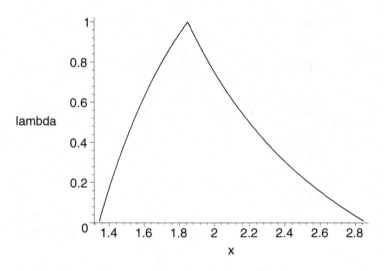

Figure 9.4: Fuzzy Estimator $\overline{\sigma}$ in Example 9.3.1, $0.01 \leq \beta \leq 1$

To complete this chapter let us present one graph of our fuzzy estimator $\overline{\sigma}$ of σ. Alpha-cuts of $\overline{\sigma}$ are

$$[\frac{9.06}{\sqrt{L(\lambda)}}, \frac{9.06}{\sqrt{R(\lambda)}}]. \tag{9.15}$$

Assuming $0.01 \leq \beta \leq 1$, the graph is in Figure 9.4.

9.4   References

1. R.V. Hogg and E.A. Tanis: Probability and Statistical Inference, Sixth Edition, Prentice Hall, Upper Saddle River, N.J., 2001.

2. Maple 9, Waterloo Maple Inc., Waterloo, Canada.

Chapter 10

Fuzzy Arrival/Service Rates

10.1 Introduction

In this chapter we concentrate on deriving fuzzy number estimators for the arrival rate, and the service rate, in a queuing system. These results have been employed in [1] and [2].

10.2 Fuzzy Arrival Rate

We assume that we have Poisson arrivals [5] which means that there is a positive constant λ so that the probability of k arrivals per unit time is

$$\lambda^k \exp(-\lambda)/k!, \tag{10.1}$$

the Poisson probability function. We need to estimate λ, the arrival rate, so we take a random sample $X_1, ..., X_m$ of size m. In the random sample X_i is the number of arrivals per unit time, in the ith observation. Let S be the sum of the X_i and let $\overline{X}$ be S/m. Here, $\overline{X}$ is not a fuzzy set but the mean.

Now S is Poisson with parameter $m\lambda$ ([3], p. 298). Assuming that $m\lambda$ is sufficiently large (say, at least 30), we may use the normal approximation ([3], p. 317), so the statistic

$$W = \frac{S - m\lambda}{\sqrt{m\lambda}}, \tag{10.2}$$

is approximately a standard normal. Then

$$P[-z_{\beta/2} < W < z_{\beta/2}] \approx 1 - \beta, \tag{10.3}$$

James J. Buckley: *Fuzzy Probability and Statistics*, StudFuzz **196**, 9 5 – 99 (2006)
www.springerlink.com © Springer-Verlag Berlin Heidelberg 2006

where the $z_{\beta/2}$ was defined in equation (6.7) in Chapter 6. Now divide numerator and denominator of W by m and we get

$$P[-z_{\beta/2} < Z < z_{\beta/2}] \approx 1 - \beta, \qquad (10.4)$$

where

$$Z = \frac{\overline{X} - \lambda}{\sqrt{\lambda/m}}. \qquad (10.5)$$

From these last two equations we may derive an approximate $(1 - \beta)100\%$ confidence interval for λ. Let us call this confidence interval $[l(\beta), r(\beta)]$.

We now show how to compute $l(\beta)$ and $r(\beta)$. Let

$$f(\lambda) = \sqrt{m}(\overline{X} - \lambda)/\sqrt{\lambda}. \qquad (10.6)$$

Now $f(\lambda)$ has the following properties: (1) it is strictly decreasing for $\lambda > 0$; (2) it is zero for $\lambda > 0$ only at $\overline{X} = \lambda$; (3) the limit of f, as λ goes to ∞ is $-\infty$; and (4) the limit of f as λ approaches zero from the right is ∞. Hence, (1) the equation $z_{\beta/2} = f(\lambda)$ has a unique solution $\lambda = l(\beta)$; and (2) the equation $-z_{\beta/2} = f(\lambda)$ also has a unique solution $\lambda = r(\beta)$.

We may find these unique solutions. Let

$$V = \sqrt{z_{\beta/2}^2/m + 4\overline{X}}, \qquad (10.7)$$

$$z_1 = [-\frac{z_{\beta/2}}{\sqrt{m}} + V]/2, \qquad (10.8)$$

and

$$z_2 = [\frac{z_{\beta/2}}{\sqrt{m}} + V]/2. \qquad (10.9)$$

Then $l(\beta) = z_1^2$ and $r(\beta) = z_2^2$.

We now substitute α for β to get the α-cuts of fuzzy number $\overline{\lambda}$. Add the point estimate, when $\alpha = 1$, $\overline{X}$, for the 0% confidence interval. Now as α goes from 0.01 (99% confidence interval) to one (0% confidence interval) we get a triangular shaped fuzzy number for λ. As before, we drop the graph straight down at the ends to obtain a complete fuzzy number.

Example 10.2.1

Suppose $m = 100$ and we obtained $\overline{X} = 25$. We evaluated equations (10.7) through (10.9) using Maple [4] and then the graph of $\overline{\lambda}$ is shown in Figure 10.1, without dropping the graph straight down to the x-axis at the end points. However, in applications, for simplicity we usually use a triangular fuzzy number for $\overline{\lambda}$. The Maple commands are in Chapter 30.

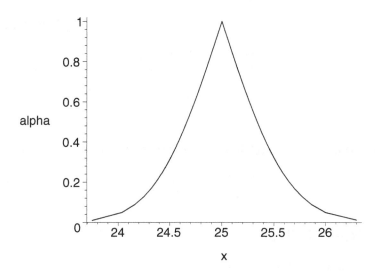

Figure 10.1: Fuzzy Arrival Rate $\overline{\lambda}$ in Example 10.2.1

10.3 Fuzzy Service Rate

Let μ be the average (expected) service rate, in the number of service completions per unit time, for a busy server. Then $1/\mu$ is the average (expected) service time. The probability density of the time interval between successive service completions is ([5], Chapter 15)

$$(1/\mu)\exp(-t/\mu), \tag{10.10}$$

for $t > 0$, the exponential probability density function. Let $X_1, ..., X_n$ be a random sample from this exponential density function. Then the maximum likelihood estimator for μ is $\overline{X}$ ([3],p.344), the mean of the random sample (not a fuzzy set). We know that the probability density for $\overline{X}$ is the gamma ([3],p.297) with mean μ and variance μ^2/n ([3],p.351). If n is sufficiently large we may use the normal approximation to determine approximate confidence intervals for μ. Let

$$Z = (\sqrt{n}[\overline{X} - \mu])/\mu, \tag{10.11}$$

which is approximately normally distributed with zero mean and unit variance, provided n is sufficiently large. See Figure 6.4-2 in [3] for $n = 100$ which shows the approximation is quite good if $n = 100$. The graph in Figure 6.4-2 in [3] is for the chi-square distribution which is a special case of the gamma distribution. So we now assume that $n \geq 100$ and use the normal approximation to the gamma.

An approximate $(1 - \beta)100\%$ confidence interval for μ is obtained from

$$P[-z_{\beta/2} < Z < z_{\beta/2}] \approx 1 - \beta, \tag{10.12}$$

where β was defined in equation (6.7) of Chapter 6. After solving for μ we get

$$P[L(\beta) < \mu < R(\beta)] \approx 1 - \beta, \tag{10.13}$$

where

$$L(\beta) = [\sqrt{n}\,\overline{X}]/[z_{\beta/2} + \sqrt{n}], \tag{10.14}$$

and

$$R(\beta) = [\sqrt{n}\,\overline{X}]/[\sqrt{n} - z_{\beta/2}]. \tag{10.15}$$

An approximate $(1 - \beta)100\%$ confidence interval for μ is

$$[\frac{\sqrt{n}\,\overline{X}}{z_{\beta/2} + \sqrt{n}}, \frac{\sqrt{n}\,\overline{X}}{\sqrt{n} - z_{\beta/2}}]. \tag{10.16}$$

Example 10.3.1

If $n = 400$ and $\overline{X} = 1.5$, then we get

$$[\frac{30}{z_{\beta/2} + 20}, \frac{30}{20 - z_{\beta/2}}], \tag{10.17}$$

for a $(1 - \beta)100\%$ confidence interval for the service rate μ. Now we can put these confidence intervals together, one on top of another, to obtain a fuzzy number $\overline{\mu}$ for the service rate. We evaluated equation (10.17) using Maple [4] for $0.05 \leq \beta \leq 1$ and the graph of the fuzzy service rate, without dropping the graph straight down to the x-axis at the end points, is in Figure 10.2. For simplicity we usually use triangular fuzzy numbers for $\overline{\mu}$ in applications. Maple commands for this figure are in Chapter 30.

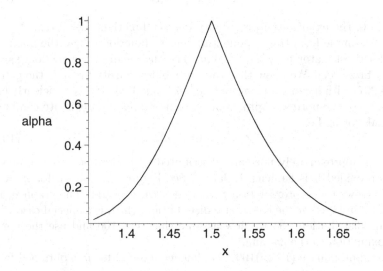

Figure 10.2: Fuzzy Service Rate $\overline{\mu}$ in Example 10.3.1

10.4 References

1. J.J. Buckley: Fuzzy Probabilities and Fuzzy Sets for Web Planning, Springer, Heidelberg, Germany, 2004.

2. J.J. Buckley: Simulating Fuzzy Systems, Springer, Heidelberg, Germany, 2005.

3. R.V. Hogg and E.A. Tanis: Probability and Statistical Inference, Sixth Edition, Prentice Hall, Upper Saddle River, N.J., 2001.

4. Maple 9, Waterloo Maple Inc., Waterloo, Canada.

5. H.A. Taha: Operations Research, Fifth Edition, Macmillan, N.Y., 1992.

Chapter 11

Fuzzy Uniform

11.1 Introduction

Let $U(a, b)$, for $0 \leq a < b$, denote the uniform distribution whose density function is $f(x) = 1/(b - a)$ for $a < x < b$, and $f(x) = 0$ otherwise. We have used the fuzzy uniform distribution to model times between arrivals in a queuing system and/or service times in a server ([1],[2]).

11.2 Fuzzy Estimators

Assume that we do not know a and b precisely so they must be estimated. Let $X_1, ..., X_n$ be a random sample from $U(a, b)$ and let $x_1, ..., x_n$ be the values of this random sample. Define $x_{min} = min\{x_i | 1 \leq i \leq n\}$ and $x_{max} = max\{x_i | 1 \leq i \leq n\}$. Then x_{min} (x_{max}) is our point estimator of a (b). We may now construct $(1 - \beta)100\%$ confidence intervals for a (b), for $0.01 \leq \beta \leq 1$, and placing these intervals one on top of another, with completing the base as described in Chapter 6, to obtain our fuzzy estimator $\overline{a}$ ($\overline{b}$) of a (b).

We will describe the mechanics of building these confidence intervals in the next section. The solution for the confidence intervals is complicated [4] and not to be found in the usual statistics books. In fact, we will first construct a joint confidence region for (a, b) and project it onto the $a - axis$ ($b - axis$) for a confidence interval for a (b).

11.2.1 Details

Let $A = a + \epsilon(b - a)$ and $B = b - \epsilon(b - a)$ for some $0 \leq \epsilon \leq 1$. Consider the event $E_1 = \{x_{min} | a \leq x_{min} \leq A\}$ and the event $E_2 = \{x_{max} | B \leq x_{max} \leq b\}$. We know that $a \leq x_{min}$ and $x_{max} \leq b$ are both always true, so it is the other inequalities $x_{min} \leq A$ and $B \leq x_{max}$ we are interested in. Now (P denotes probability, E^c denotes compliment) a joint confidence region for (a, b) is

James J. Buckley: *Fuzzy Probability and Statistics*, StudFuzz **196**, $101-105$ (2006)
www.springerlink.com

defined by

$$P(E_1 \cap E_2) = 1 - \beta, \tag{11.1}$$

for $0.01 \leq \beta \leq 1$. We see that

$$P(E_1 \cap E_2) = 1 - P(E_1^c \cup E_2^c) = 1 - P(E_1^c) - P(E_2^c) + P(E_1^c \cap E_2^c) = \tag{11.2}$$

$$1 - P(x_{min} > A) - P(x_{max} < B) + P(A < \text{all } x_i's < B) = \tag{11.3}$$

$$1 - [(b - A)/(b - a)]^n - [(B - a)/(b - a)]^n + [(B - A)/(b - a)]^n = \tag{11.4}$$

$$1 - (1 - \epsilon)^n - (1 - \epsilon)^n + (1 - 2\epsilon)^n = 1 - \beta, \tag{11.5}$$

or

$$2(1 - \epsilon)^n - (1 - 2\epsilon)^n = 1 - \beta. \tag{11.6}$$

Given β we solve the last equation for ϵ which is used in the definitions of A and B.

Let $f(\epsilon) = 2(1 - \epsilon)^n - (1 - 2\epsilon)^n - 1 + \beta$ for $0 \leq \epsilon \leq 1$. As β increases from 0.01 to 1 we solve $f(\epsilon) = 0$ for ϵ, construct A and B and the joint confidence region for (a, b). Solutions for selected values of n and for $\beta = 0.01, 0.50$ are shown in Tables 11.1 and 11.2, respectively. We have two situations, determined from the graph of $f(\epsilon)$, which are: (1) if n is even the is a unique solution for ϵ in $[0, 1]$; (2) if n is odd there are two solutions for ϵ in $[0, 1]$ and we always picked the smaller value. Solutions and graphs were done using Maple [3].

The joint confidence region for (a, b) is defined by the following four inequalities:

$$a \leq x_{min}, \tag{11.7}$$

$$x_{min} \leq A = a + \epsilon(b - a), \tag{11.8}$$

$$x_{max} \leq b, \tag{11.9}$$

$$B = b - \epsilon(b - a) \leq x_{max}. \tag{11.10}$$

n	ϵ
20	0.23272
21	0.22297
30	0.16187
31	0.15708
50	0.10052
51	0.09865

Table 11.1: Values of ϵ for Various Values of n, Given $\beta = 0.01$

n	ϵ
20	0.06030
21	0.05748
30	0.04044
31	0.03915
50	0.02438
51	0.02391

Table 11.2: Values of ϵ for Various Values of n, Given $\beta = 0.50$

Example 11.2.1

Let $a = 1$, $b = 10$, $n = 20$, and $\beta = 0.05$ for 95% confidence intervals for a and b. Assume that we obtained $x_{min} = 2$ and $x_{max} = 8$. Now construct a coordinate system with the a axis horizontal and the b axis vertical as in Figure 11.1. We solve for ϵ using Maple and get $\epsilon = 0.16821$. First put in the graph of $A = (1 - \epsilon)a + \epsilon b = 2 = x_{min}$ giving a-intercept 2.4 and b-intercept 11.9. Next make the graph of $B = \epsilon a + (1 - \epsilon)b = 8 = x_{max}$ with a-intercept 47.6 and b-intercept 9.6. The confidence region is below B, above A, to the

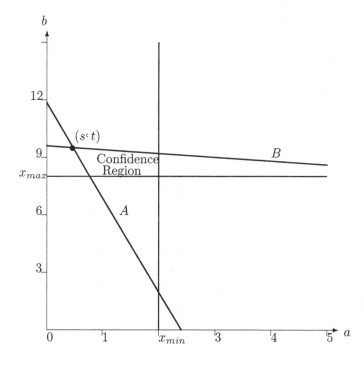

Figure 11.1: Determining Joint Confidence Region for (a, b)

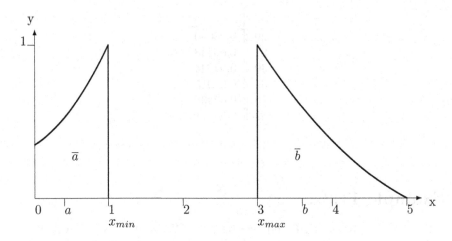

Figure 11.2: Fuzzy Estimators $\overline{a}$ and $\overline{b}$ in $U(a,b)$

left of x_{min} and above x_{max}. We computed $(s,t) = (0.479, 9.521)$ where (s,t) is shown in Figure 11.1. Now project the confidence region onto the a-axis and we get $[0.479, 2]$ as our 95% confidence interval for a. Next project onto the b-axis and we obtain $[8, 9.521]$ as the 95% confidence interval for b.

Do this for β increasing from 0.01 to 1 producing a family of confidence intervals. We place these intervals one on top of another as described in Chapter 6 generating half triangular shaped fuzzy number estimators $\overline{a}$ and $\overline{b}$ for a and b, respectively. Then we complete the bottoms of these fuzzy numbers as discussed in Chapter 6. The bottoms of the fuzzy estimators are 99% confidence intervals.

Typical fuzzy estimators $\overline{a}$ and $\overline{b}$ are shown in Figures 11.2 and 11.3. In Figure 11.2 the left side of $\overline{a}$ was cut off as we assume that a is non-negative. These graphs were done using LaTeX .

If we go back to the fuzzy uniform distribution in Section 5.2 in Chapter 5 we see that there we used (whole) triangular fuzzy number estimators for both a and b. This is not correct since the $\overline{a}$ and $\overline{b}$ in Figures 11.2 and 11.3 are not complete triangular shaped fuzzy numbers. They are half of a triangular shaped fuzzy number. It is just a habit of the author to use triangular fuzzy numbers in applications. So go back to Section 5.2 and change those triangular fuzzy estimators to be a half of a triangular fuzzy number. The left half for $\overline{a}$ and the right half for $\overline{b}$. We leave this as an exercise for the interested reader. Changing $\overline{a}$ and $\overline{b}$ in Section 5.2 to look like those in Figure 11.3 will probably only cause small changes in the results. Is this true?

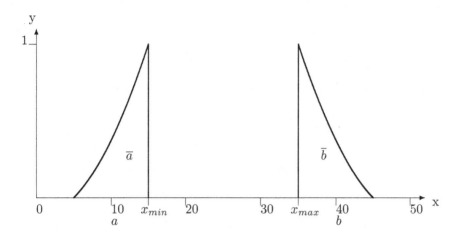

Figure 11.3: Fuzzy Estimators $\overline{a}$ and $\overline{b}$ in the Uniform Distribution

11.3 References

1. J.J. Buckley: Fuzzy Probabilities and Fuzzy Sets for Web Planning, Springer, Heidelberg, Germany, 2004.

2. J.J. Buckley: Simulating Fuzzy Systems, Springer, Heidelberg, Germany, 2005.

3. Maple 9, Waterloo Maple Inc., Waterloo, Canada.

4. Personal Communication from Professor N.Chernov, Mathematics Department, UAB, Birmingham, Alabama.

Chapter 12

Fuzzy Max Entropy Principle

12.1 Introduction

We first discuss the maximum entropy principle, subject to crisp (non-fuzzy) constraints, in the next section. This presentation is based on [1]. Then we show how this principle may be extended to handle fuzzy constraints (fuzzy numbers model the imprecision) in Section 12.3. In Section 12.3 we obtain solutions like a fuzzy discrete probability distribution, the fuzzy normal probability distribution, the fuzzy negative exponential distribution, etc. which were all discussed in Chapters 4 and 5. These results are modeled after [2]. The following two chapters continues our work with the fuzzy maximum entropy principle.

12.2 Maximum Entropy Principle

We first consider discrete probability distributions and then continuous probability distributions. The entropy principle has not gone uncriticized, and this literature, together with that justifying the principle, has been surveyed in [1].

12.2.1 Discrete Probability Distributions

We start with a discrete, and finite, probability distribution. Let $X = \{x_1, ..., x_n\}$ and $p_i = P(x_i)$, $1 \leq i \leq n$, where we use P for probability. We do not know all the p_i values exactly but we do have some prior information, possibly through expert opinion, about the distribution. This information could be in the form of: (1) its mean; (2) its variance; or (3) interval estimates for the p_i. The decision problem is to find the "best" $p = (p_1, ..., p_n)$

James J. Buckley: *Fuzzy Probability and Statistics*, StudFuzz **196**, 107–114 (2006)
 © Springer-Verlag Berlin Heidelberg 2006

subject to the constraints given in the information we have about the distribution. A measure of uncertainty in our decision problem is computed by $H(p) = H(p_1, ..., p_n)$ where

$$H(p) = -\sum_{i=1}^{n} p_i \ln(p_i), \qquad (12.1)$$

for $p_1 + ... + p_n = 1$ and $p_i \geq 0$, $1 \leq i \leq n$. Define $0 \ln(0) = 0$. $H(p)$ is called the entropy (uncertainty) in the decision problem.

Let $\mathcal{F}$ denote the set of feasible probability vectors p. $\mathcal{F}$ will contain all the p satisfying the constraints dictated by the prior information about the distribution. The maximum entropy principle states that the "best" p, say p^*, has the maximum entropy subject to $p \in \mathcal{F}$. Therefore p^* solves

$$max[-\sum_{i=1}^{n} p_i \ln(p_i)], \qquad (12.2)$$

subject to $p \in \mathcal{F}$. With only the constraint that $p_1 + ... + p_n = 1$ and $p_i \geq 0$ all i the solution is the uniform distribution $p_i = 1/n$ all i.

It is easy to extend this decision problem to the infinite case of $X = \{x_1, ..., x_n, ...\}$.

Example 12.2.1.1

Suppose we have prior information, possibly through expert opinion, about the mean m of the discrete probability distribution. Our decision problem is

$$max[-\sum_{i=1}^{n} p_i \ln(p_i)], \qquad (12.3)$$

subject to

$$p_1 + ... + p_n = 1, \; p_i \geq 0, \; 1 \leq i \leq n, \qquad (12.4)$$

$$\sum_{i=1}^{n} x_i p_i = m. \qquad (12.5)$$

The solution is [1]

$$p_i^* = \exp[\lambda - 1] \exp[\mu x_i], \qquad (12.6)$$

for $1 \leq i \leq n$ and λ and μ are Lagrange multipliers whose values are obtained from the constraints

$$\exp[\lambda - 1] \sum_{i=1}^{n} \exp[\mu x_i] = 1, \qquad (12.7)$$

$$\exp[\lambda - 1] \sum_{i=1}^{n} x_i \exp[\mu x_i] = m. \qquad (12.8)$$

An example where the constraints are $p_1 + ... + p_n = 1$, $p_i \geq 0$ all i and $a_i \leq p_i \leq b_i$ all i with $a_1 + ... + a_n \leq 1 \leq b_1 + ... + b_n$ is in [1].

Example 12.2.1.2

Now assume that $X = \{0, 1, 2, 3, ...\}$ so that we have a discrete, but infinite, probability distribution. If we have prior information about the expected outcome m, then the decision problem is

$$max[-\sum_{i=0}^{\infty} p_i \ln(p_i)], \qquad (12.9)$$

subject to

$$\sum_{i=0}^{\infty} p_i = 1, \; p_i \geq 0, \; all \; i, \qquad (12.10)$$

$$\sum_{i=0}^{\infty} ip_i = m. \qquad (12.11)$$

The solution, using Lagrange multipliers, is [1]

$$p_i^* = (\frac{1}{m+1})(\frac{m}{m+1})^i, \; i = 0, 1, 2, 3, ... \qquad (12.12)$$

which is the geometric probability distribution.

12.2.2 Continuous Probability Distributions

Let E be (a, b), $-\infty < a < b < \infty$, or $(0, \infty)$, or $(-\infty, \infty)$. The probability density function over E will be written as $f(x)$. That is, $f(x) \geq 0$ for $x \in E$ and $f(x)$ is zero outside E. We do not know the probability density function exactly but we do have some prior information, possibly through expert opinion, about the distribution. This information could be in the form of: (1) its mean; or (2) its variance. The decision problem is to find the "best" $f(x)$ subject to the constraints given in the information we have about the distribution. A measure of uncertainty (entropy) in our decision problem is $H(f(x))$ computed by

$$H(f(x)) = -\int_E f(x) \ln[f(x)]dx, \qquad (12.13)$$

for $f(x) \geq 0$ on E and the integral of $f(x)$ over E equals one. Define $0 \ln(0) = 0$. $H(f(x))$ is called the entropy (uncertainty) in the decision problem.

Let $\mathcal{F}$ denote the set of feasible probability density functions. $\mathcal{F}$ will contain all the $f(x)$ satisfying the constraints dictated by the prior information about the distribution. The maximum entropy principle states that the "best" $f(x)$, say $f^*(x)$, has the maximum entropy subject to $f(x) \in \mathcal{F}$. Therefore $f^*(x)$ solves

$$max[-\int_E f(x)\ln[f(x)]dx], \qquad (12.14)$$

subject to $f(x) \in \mathcal{F}$. With only the constraint that $\int_E f(x)dx = 1$, $f(x) \geq 0$ on E and $E = (a, b)$ the solution is the uniform distribution on E.

Example 12.2.2.1

Suppose we have prior information, possibly through expert opinions, about the mean m and variance σ^2 of the probability density. Our decision problem is

$$max\{-\int_E f(x)\ln[f(x)]dx\}, \qquad (12.15)$$

subject to

$$\int_E f(x)dx = 1, \ f(x) \geq 0 \ on \ E, \qquad (12.16)$$

$$\int_E xf(x)dx = m, \qquad (12.17)$$

$$\int_E (x - m)^2 f(x)dx = \sigma^2. \qquad (12.18)$$

The solution, using the calculus of variations, is [1]

$$f^*(x) = \exp[\lambda - 1]\exp[\mu x]\exp[\gamma(x - m)^2], \qquad (12.19)$$

where the constants λ, μ, γ are determined from the constraints given in equations (12.16) through (12.18).

Example 12.2.2.2

Let $E = (0, \infty)$ and omit the constraint that the variance must equal the positive number σ^2. That is, in Example 12.2.2.1 drop the constraint in equation (12.18). Then the solution is [1]

$$f^*(x) = (1/m)\exp[-\frac{x}{m}], \ x \geq 0, \qquad (12.20)$$

the negative exponential.

Example 12.2.2.3

Now assume that $E = (-\infty, \infty)$ together with all the constraints of Example 12.2.2.1. The solution is [1] the normal probability density with mean m and variance σ^2.

12.3 Imprecise Side-Conditions

We first consider discrete probability distributions and then continuous probability distributions. We will only consider imprecise side-conditions relating to the mean and variance of the unknown probability distribution. These imprecise conditions will be stated as the mean is "approximately" m and the variance is "approximately" σ^2. We will model this imprecision using triangular fuzzy numbers.

How will we obtain these fuzzy numbers? One way is through expert opinion discussed in Section 3.3 or from data in Chapters 6,7 and 9. We will use the triangular fuzzy number $\overline{m} = (m_1/m_2/m_3)$ for "approximately" m. Similarly we use $\overline{\sigma}^2 = (\sigma_1^2/\sigma_2^2/\sigma_3^2)$ with $\sigma_1^2 > 0$ for $\approx \sigma^2$.

We now show how to solve the maximum entropy principle with imprecise side-conditions through a series of examples patterned after the examples in the previous section.

12.3.1 Discrete Probability Distributions

Example 12.3.1.1

This is the same as Example 12.2.1.1 except equation (12.5) becomes

$$\sum_{i=1}^{n} x_i p_i = \overline{m}. \tag{12.21}$$

We solve by taking α-cuts. So the above equation becomes

$$\sum_{i=1}^{n} x_i p_i = \overline{m}[\alpha], \tag{12.22}$$

for $\alpha \in [0,1]$. Now we solve the decision problem, equations (12.3), (12.4) and (12.22), for each $m \in \overline{m}[\alpha]$ giving

$$\overline{\Omega}[\alpha] = \{p^* \mid m \in \overline{m}[\alpha]\}, \tag{12.23}$$

for each α. We put these α-cuts together to obtain the fuzzy set $\overline{\Omega}$, a fuzzy subset of $\mathbf{R}^n$.

We can not project the joint fuzzy probability distribution $\overline{\Omega}$ onto the coordinate axes to get the marginal fuzzy probabilities because the α-cuts of $\overline{\Omega}$ are not "rectangles" in $\mathbf{R}^n$. In fact, $\overline{\Omega}$ is a fuzzy subset of the hyperplane $\{p = (p_1, ..., p_n) | p_1 + ... + p_n = 1\}$.

How can we compute fuzzy probabilities using $\overline{\Omega}$? The basic method is like our restricted fuzzy arithmetic contained in Chapter 3. Let A be a subset on X. Say $A = \{x_1, x_2, ..., x_6\}$. We want $\overline{P}(A)$ the fuzzy probability of A. It is to be determined by its α-cuts

$$\overline{P}(A)[\alpha] = \{p_1 + ... + p_6 \mid p \in \overline{\Omega}[\alpha]\}, \tag{12.24}$$

for all α. Now this α-cut will be an interval so let $\overline{P}(A)[\alpha] = [\tau_1(\alpha), \tau_2(\alpha)]$. Then the optimization problems give the end points of this interval

$$\tau_1(\alpha) = min\{p_1 + ... + p_6 \mid p \in \overline{\Omega}[\alpha]\}, \tag{12.25}$$

$$\tau_2(\alpha) = max\{p_1 + ... + p_6 \mid p \in \overline{\Omega}[\alpha]\}, \tag{12.26}$$

all α.

Next we might ask is the mean of $\overline{\Omega}$ equal to $\overline{m}$. We now see if this is true. The fuzzy mean is computed by α-cuts. Let this unknown fuzzy mean be $\overline{M}$. Then

$$\overline{M}[\alpha] = \{ \sum_{i=1}^{n} x_i p_i \mid p \in \overline{\Omega}[\alpha] \}, \tag{12.27}$$

all α. But each $p \in \overline{\Omega}[\alpha]$ corresponds to a $m \in \overline{m}[\alpha]$ so the sum in equation (12.27) equals the m that produced the p we chose in $\overline{\Omega}[\alpha]$. Hence, $\overline{M}[\alpha] = \overline{m}[\alpha]$ all α and $\overline{M} = \overline{m}$.

Example 12.3.1.2

This is the same as Example 12.2.1.2 except equation (12.11) is

$$\sum_{i=0}^{\infty} i p_i = \overline{m}. \tag{12.28}$$

As in the previous example we solve by α-cuts producing $\overline{\Omega}[\alpha]$ and $\overline{\Omega}$.

It is easier to see what we get in this case because the $p \in \overline{\Omega}[\alpha]$ are given by equation (12.12) for all $m \in \overline{m}[\alpha]$. We again may find that the fuzzy mean of $\overline{\Omega}$ is $\overline{m}$.

12.3.2   Continuous Probability Distributions

Example 12.3.2.1

This example continues Example 12.2.2.1 but now we have fuzzy mean $\overline{m}$ and fuzzy variance $\overline{\sigma}^2$. We solve by α-cuts. That is, we solve the optimization problem in Example 12.2.2.1 for all m in $\overline{m}[\alpha]$ and for all σ^2 in $\overline{\sigma}^2[\alpha]$. This produces $\overline{\Omega}[\alpha]$ and $\overline{\Omega}$. That is

$$\overline{\Omega}[\alpha] = \{f^*(x) \mid m \in \overline{m}[\alpha], \ \sigma^2 \in \overline{\sigma}^2[\alpha] \}. \tag{12.29}$$

How do we compute fuzzy probabilities with this joint fuzzy distribution? Let G be a subset of E. Then an α-cut of $\overline{P}(G)$ is

$$\overline{P}(G)[\alpha] = \{\int_G f^*(x)dx \mid m \in \overline{m}[\alpha], \ \sigma^2 \in \overline{\sigma}^2[\alpha] \}, \tag{12.30}$$

for all α. $\overline{P}(G)$ is a fuzzy subset of $\mathbf{R}$ and its interval α-cuts are given in the above equation.

We may also find the fuzzy mean and fuzzy variance of $\overline{\Omega}$ and compute $\overline{m}$ and $\overline{\sigma}^2$, respectively. For example, if we denote the fuzzy mean of $\overline{\Omega}$ as $\overline{M}$ its alpha-cuts are

$$\overline{M}[\alpha] = \{ \int_E x f^*(x) dx \mid m \in \overline{m}[\alpha], \ \sigma^2 \in \overline{\sigma}^2[\alpha] \}, \tag{12.31}$$

for all alpha. Now the integral in the above equation equals m for each m in the alpha-cut of $\overline{m}$ and σ^2 in the alpha-cut of $\overline{\sigma}^2$. So $\overline{M}[\alpha] = \overline{m}[\alpha]$ all alpha and $\overline{M} = \overline{m}$.

Example 12.3.2.2

This is the same as Example 12.2.2.2 but it has a fuzzy mean $\overline{m}$. Solving by α-cuts we obtain the fuzzy negative exponential.

Example 12.3.2.3

The same as Example 12.2.2.3 having a fuzzy mean and a fuzzy variance. Solving by α-cuts we get the fuzzy normal with mean $\overline{m}$ and variance $\overline{\sigma}^2$.

Let $N(c, d)$ denote the normal probability density with mean c and variance d. Then

$$\overline{\Omega}[\alpha] = \{ N(m, \sigma^2) \mid m \in \overline{m}[\alpha], \ \sigma^2 \in \overline{\sigma}^2[\alpha] \}, \tag{12.32}$$

for $\alpha \in [0, 1]$. We compute with the fuzzy normal as follows

$$\overline{P}(G)[\alpha] = \{ \int_G N(m, \sigma^2) dx \mid m \in \overline{m}[\alpha], \ \sigma^2 \in \overline{\sigma}^2[\alpha] \}, \tag{12.33}$$

for all alpha giving fuzzy probability $\overline{P}(G)$. We may also find that the fuzzy mean of $\overline{\Omega}$ is $\overline{m}$ and the fuzzy variance of $\overline{\Omega}$ is $\overline{\sigma}^2$.

12.4 Summary and Conclusions

We solved the maximum entropy principle with imprecise side-conditions, which were modeled as fuzzy sets, producing fuzzy probability distributions. It seems very natural if you start with a fuzzy mean, variance, etc, you need to end up with a fuzzy probability distribution. Fuzzy probability distributions produce fuzzy means, variances, etc. In the next two chapters we restrict the solutions to be crisp (not fuzzy).

12.5   References

1. J.J. Buckley: Entropy Principles in Decision Making Under Risk, Risk Analysis, 5(1985)303-313.

2. J.J. Buckley: Maximum   Entropy   Principle   with   Imprecise   Side-Conditions, Soft Computing, 9(2005)507-511.

Chapter 13

Max Entropy: Crisp Discrete Solutions

13.1   Introduction

We first discuss the maximum entropy principle, subject to crisp (non-fuzzy) constraints, in the next section. This was presented in the previous chapter but for completeness we repeat some of that discussion again in this chapter. This presentation is restricted to discrete probability distributions and is based on [2]. Then we show how this principle may be extended to handle fuzzy constraints (fuzzy numbers model the imprecision) in Section 13.3. In Section 13.3 we obtain crisp solutions to the discrete probability case. These results are based on [3]. In the previous chapter we obtained solutions like fuzzy discrete probability distributions, the fuzzy normal probability distribution, and the fuzzy negative exponential distribution to the maximum entropy problem.

13.2   Max Entropy: Discrete Distributions

The entropy principle has not gone uncriticized, and this literature, together with that justifying the principle, has been surveyed in [2].

We have a discrete, and finite, probability distribution. Let $X = \{x_1, ..., x_n\}$ and $p_i = P(x_i)$, $1 \leq i \leq n$, where we use P for probability. We do not know all the p_i values exactly but we do have some prior information, possibly through expert opinion, about the distribution. This information could be in the form of: (1) its mean; (2) its variance; or (3) interval estimates for the p_i. The decision problem is to find the "best" $p = (p_1, ..., p_n)$ subject to the constraints given in the information we have about the distribution. A measure of uncertainty in our decision problem is computed by

James J. Buckley: *Fuzzy Probability and Statistics*, StudFuzz **196**, 115–124 (2006)
www.springerlink.com   © Springer-Verlag Berlin Heidelberg 2006

$F(p) = F(p_1, ..., p_n)$ where

$$F(p) = -\sum_{i=1}^{n} p_i \ln(p_i),\tag{13.1}$$

for $p_1 + ... + p_n = 1$ and $p_i \geq 0$, $1 \leq i \leq n$. Define $0 \ln(0) = 0$. $F(p)$ is called the entropy (uncertainty) in the decision problem.

Let $\mathcal{F}$ denote the set of feasible probability vectors p. $\mathcal{F}$ will contain all the p satisfying the constraints dictated by the prior information about the distribution. The maximum entropy principle states that the "best" p, say p^*, has the maximum entropy subject to $p \in \mathcal{F}$. Therefore p^* solves

$$max[-\sum_{i=1}^{n} p_i \ln(p_i)],\tag{13.2}$$

subject to $p \in \mathcal{F}$. With only the constraint that $p_1 + ... + p_n = 1$ and $p_i \geq 0$ for all i, the solution is the uniform distribution $p_i = 1/n$ for all i.

Example 13.2.1

Suppose we have prior information, possibly through expert opinion, about the mean m of the discrete probability distribution and its variance σ^2. Our decision problem is

$$max[-\sum_{i=1}^{n} p_i \ln(p_i)],\tag{13.3}$$

subject to

$$p_1 + ... + p_n = 1, \ p_i \geq 0, \ 1 \leq i \leq n,\tag{13.4}$$

$$\sum_{i=1}^{n} x_i p_i = m,\tag{13.5}$$

$$\sum_{i=1}^{n} (x_i - m)^2 p_i = \sigma^2.\tag{13.6}$$

13.3   Max Entropy: Imprecise Side-Conditions

We will only consider imprecise side-conditions relating to the mean and variance of the unknown probability distribution. These imprecise conditions will be stated as the mean is "approximately" m and the variance is "approximately" σ^2. We will model this imprecision using triangular fuzzy numbers. How will we obtain these fuzzy numbers? From expert opinion (Section 3.3)and/or from data (Chapters 6-9). We will use the triangular fuzzy number $\overline{m} = (m_1/m_2/m_3)$ for "approximately" m. Similarly we use $\overline{\sigma}^2 = (\sigma_1^2/\sigma_2^2/\sigma_3^2)$ with $\sigma_1^2 > 0$ for $\approx \sigma^2$.

We now show how to solve the maximum entropy principle with imprecise side-conditions through a series of examples. We first assume that the only prior information is about the mean and we solve both problems: mean=m and mean=$\overline{m}$. Then we assume that the prior information extends to also include the variance and solve both problems. All optimization problems are solved using "SOLVER" in Excel ([4],[6]). Further information on SOLVER and the programming of SOLVER is contained in Chapter 30. All examples use the data $x_1 = 0, x_2 = 1, ..., x_5 = 4$. We can easily expand to cover larger problems (more x_i values).

Example 13.3.1

We want to find the p_i, $1 \leq i \leq 5$, which solve

$$max[-\sum_{i=1}^{5} p_i \ln(p_i)], \tag{13.7}$$

subject to

$$p_1 + ... + p_5 = 1, \ p_i \geq 0, \ 1 \leq i \leq 5, \tag{13.8}$$

$$\sum_{i=1}^{5} x_i p_i = 3. \tag{13.9}$$

The prior information was that the mean equals 3. This is easily solved using SOLVER and the results are in Table 13.1. More details on how Solver was used to solve this optimization problem is in Chapter 30. One may also solve this optimization problem by solving a system of two nonlinear equations simultaneously, given in [2], and we did this using Maple [5], with the results agreeing with those in Table 13.1. Also, we solved this crisp problem only to contrast with the fuzzy problem solved next.

x	$P(x)$
0	0.0477
1	0.0841
2	0.1481
3	0.2608
4	0.4594

Table 13.1: Solution to Crisp Maximum Entropy in Example 13.3.1

The value of the mean for the probability distribution in Table 13.1 is 3 and the maximum value of the objective function (equation (13.7)) was 0.8351.

Example 13.3.2

Now we want to find the p_i, $1 \leq i \leq 5$, which solve

$$max[-\sum_{i=1}^{5} p_i \ln(p_i)], \qquad (13.10)$$

subject to

$$p_1 + ... + p_5 = 1, \; p_i \geq 0, \; 1 \leq i \leq 5, \qquad (13.11)$$

$$\sum_{i=1}^{5} x_i p_i = \overline{3}. \qquad (13.12)$$

The prior information was that the mean should equal approximately three which we model as fuzzy three $\overline{3} = (2/3/4)$, represented as the triangular fuzzy number $(2/3/4)$. The problem has changed to a fuzzy optimization problem and we will set up fuzzy goals for each objective and follow the method given in [1]. We need to construct a fuzzy goal for equation (13.10) and also for equation (13.12).

First consider equation (13.10). The function $F(p) = -\sum_{i=1}^{5} p_i \ln(p_i)$, where $p = (p_1, ..., p_5)$, varies from zero to $\ln(5)$ and attains its maximum value of $\ln(5)$ for $p_i = 1/5$ for all i. A fuzzy goal $\overline{G}(p)$ for this objective is shown in Figure 13.1. The membership function for this fuzzy goal is $\overline{G}(p) = F(p)/\ln(5)$ for $0 \leq F(p) \leq \ln(5)$ and $\overline{G}(p) = 1$ for $F(p) \geq \ln(5)$.

Next we need to get a fuzzy goal for equation (13.12). Let

$$M(p) = \sum_{i=1}^{5} x_i p_i, \qquad (13.13)$$

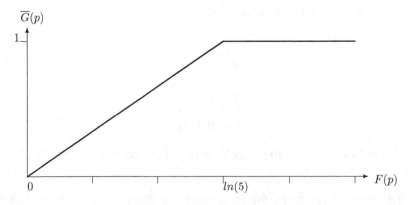

Figure 13.1: Fuzzy Goal $\overline{G}(p)$ for Example 13.3.2

which is the mean of the discrete probability distribution. For notational convenience also let $\overline{A}$ denote fuzzy three $\overline{3} = (2/3/4)$. Then the membership function for the second fuzzy goal is $\overline{A}(p) = \overline{A}(M(p))$. Then the fuzzy optimization problem becomes

$$max\{min(\overline{G}(p), \overline{A}(p))\}, \tag{13.14}$$

subject to

$$p_1 + ... + p_5 = 1, 0 \le p_i \le 1, i = 1, ..., 5. \tag{13.15}$$

We intersected the two fuzzy goals, denoted by the "min" and then we want the p of maximum membership in $\overline{G}(p) \cap \overline{A}(p)$, subject to the crisp constraints. The constraints are crisp and there is no need to fuzzify them.

To solve using SOLVER we want all functions/expressions to be "smooth" (continuous derivatives) so we need to eliminate max and min. An equivalent problem to solve, without max and min, is

$$max(X) \tag{13.16}$$

subject to

$$p_1 + ... + p_5 = 1, 0 \le p_i \le 1, i = 1, ..., 5, \tag{13.17}$$

$$X \le \overline{G}(p), X \le \overline{A}(p). \tag{13.18}$$

Finally, we need to take care of the triangular fuzzy number $\overline{A}$. We will do this by making two subproblems, solving each using SOLVER, and putting the results together to have the overall solution. Looking at the graph of $\overline{A} = (2/3/4)$ we see that the left side of the triangle is the graph of $y = x - 2$ and the right side is described by $y = 4 - x$. Then the first subproblem #1 is

$$max(X) \tag{13.19}$$

subject to

$$p_1 + ... + p_5 = 1, 0 \le p_i \le 1, i = 1, ..., 5, \tag{13.20}$$

$$X \le \overline{G}(p), X \le M(p) - 2, 2 \le M(p) \le 3, \tag{13.21}$$

and the second subproblem #2 is

$$max(X) \tag{13.22}$$

subject to

$$p_1 + ... + p_5 = 1, 0 \le p_i \le 1, i = 1, ..., 5, \tag{13.23}$$

$$X \le \overline{G}(p), X \le 4 - M(p), 3 \le M(p) \le 4. \tag{13.24}$$

Both subproblems were solved by SOLVER and the overall results are in Table 13.2. More details on SOLVER is in Chapter 30.

For subproblem #1, $2 \le M(p) \le 3$, the optimal solution was: (1) $F(p) = 0.8755$; (2) $M(p) = 2.8755$ with $\overline{A}(p) = 0.8755$; and (3) $max(X) = 0.8755$. In

x	$P(x)$
0	0.0615
1	0.0994
2	0.1605
3	0.2594
4	0.4192

Table 13.2: Solution to Fuzzy Maximum Entropy in Example 13.3.2

subproblem #2, $3 \leq M(p) \leq 4$, the optimal solution was: (1) $F(p) = 0.8351$; (2) $M(p) = 3$ so that $\overline{A}(p) = 1$; and (3) $max(X) = 0.8351$. The best of the two subproblems was #1 whose values for the p_i are given in Table 13.2.

In comparing these results to Example 13.3.1 we see $F(p)$ increased from 0.8351 to 0.8755 because we relaxed the crisp constraint that the mean should be exactly equal 3 to the mean should be approximately (modeled as a fuzzy number) equal to 3. In the fuzzy problem the optimal value of the mean ends up equal to 2.8755.

Example 13.3.3

Now assume that prior information about the discrete probability distribution encompasses both the mean and the variance. So we want to find the p_i, $1 \leq i \leq 5$, which solve

$$max[-\sum_{i=1}^{5} p_i \ln(p_i)], \qquad (13.25)$$

subject to

$$p_1 + ... + p_5 = 1, \ p_i \geq 0, \ 1 \leq i \leq 5, \qquad (13.26)$$

$$M(p) = \sum_{i=1}^{5} x_i p_i = 2, \qquad (13.27)$$

$$\sigma^2(p) = \sum_{i=1}^{5}(x_i - M(p))^2 p_i = 1. \qquad (13.28)$$

The prior information was that the mean equals 2, a change from Example 13.3.1, and the variance is one. This is easily solved using SOLVER and the results are in Table 13.3. One may also solve this optimization problem by solving a system of three nonlinear equations simultaneously, a direct extension of the two nonlinear equations used in Example 13.3.1, and we did this using Maple [5], with the results the same as in Table 13.3. More details on how Solver was used to solve this optimization problem is in Chapter 30.

x	$P(x)$
0	0.0638
1	0.2447
2	0.3830
3	0.2447
4	0.0638

Table 13.3: Solution to Crisp Maximum Entropy in Example 13.3.3

Also, we solved this crisp problem only to contrast with the fuzzy problem solved next.

The value of the mean for the probability distribution in Table 13.3 is 2, the variance equals 1, and the maximum value of the objective function (equation (13.25)) was 0.8747.

Example 13.3.4

The imprecise side-conditions are that the mean is approximately 2 and the variance is approximately one. $\overline{A} = (1/2/3)$ is the fuzzy 2 for the mean and $\overline{B} = (0/1/2)$ a fuzzy "one" for the variance. As in Example 13.3.2 $\overline{A}(p) = \overline{A}(M(p))$ and let $\overline{B}(p) = \overline{B}(\sigma^2(p))$.

Following Example 13.3.2 our fuzzy maximum entropy problem is now to solve

$$max\{min(\overline{G}(p), \overline{A}(p), \overline{B}(p))\}, \qquad (13.29)$$

subject to

$$p_1 + \ldots + p_5 = 1, 0 \le p_i \le 1, i = 1, \ldots, 5. \qquad (13.30)$$

We intersected the three fuzzy goals, denoted by the "min" and then we want the p of maximum membership in $\overline{G}(p) \cap \overline{A}(p) \cap \overline{B}(p)$, subject to the crisp constraints. The constraints are crisp and we will not fuzzify them.

An equivalent problem to solve, without max and min, is

$$max(X) \qquad (13.31)$$

subject to

$$p_1 + \ldots + p_5 = 1, 0 \le p_i \le 1, i = 1, \ldots, 5, \qquad (13.32)$$

$$X \le \overline{G}(p), X \le \overline{A}(p), X \le \overline{B}(p). \qquad (13.33)$$

Lastly, we need to take care of the triangular fuzzy numbers $\overline{A}$ and $\overline{B}$. We will do this by making four subproblems, solving each using SOLVER, and putting the results together to have the overall solution. Looking at the graph of $\overline{A} = (1/2/3)$ we see that the left side of the triangle is the graph of $y = x - 1$ and the right side is described by $y = 3 - x$. For $\overline{B} = (0/1/2)$ the left side is $y = x$ and the right side is described by equation $y = 2 - x$.

Now consider a xy-coordinate system where we label the horizontal axis $M(p)$ and the vertical axis $\sigma^2(p)$. The base of $\overline{A}$ sits on the interval $[1, 3]$ on the horizontal axis and the base of $\overline{B}$ is on the interval $[0, 2]$ on the vertical axis. We partition up the region $[1, 3] \times [0, 2]$ into four subregions as follows: (1) region #1 is $[2, 3] \times [1, 2]$; (2) region #2 is $[1, 2] \times [1, 2]$; (3) region #3 is $[1, 2] \times [0, 1]$; and (4) region #4 is $[2, 3] \times [0, 1]$. Each region corresponds to a subproblem.

Subproblem #1 is

$$max(X) \tag{13.34}$$

subject to

$$p_1 + ... + p_5 = 1, 0 \le p_i \le 1, i = 1, ..., 5, \tag{13.35}$$

$$X \le \overline{G}(p), X \le 3 - M(p), 2 \le M(p) \le 3, X \le 2 - \sigma^2(p), 1 \le \sigma^2(p) \le 2, \tag{13.36}$$

and subproblem #2 is

$$max(X) \tag{13.37}$$

subject to

$$p_1 + ... + p_5 = 1, 0 \le p_i \le 1, i = 1, ..., 5, \tag{13.38}$$

$$X \le \overline{G}(p), X \le M(p) - 1, 1 \le M(p) \le 2, X \le 2 - \sigma^2(p), 1 \le \sigma^2(p) \le 2. \tag{13.39}$$

Continuing this way subproblem #3, corresponding to region #3, is

$$max(X) \tag{13.40}$$

subject to

$$p_1 + ... + p_5 = 1, 0 \le p_i \le 1, i = 1, ..., 5, \tag{13.41}$$

$$X \le \overline{G}(p), X \le M(p) - 1, 1 \le M(p) \le 2, X \le \sigma^2(p), 0 \le \sigma^2(p) \le 1, \tag{13.42}$$

and subproblem #4 is

$$max(X) \tag{13.43}$$

subject to

$$p_1 + ... + p_5 = 1, 0 \le p_i \le 1, i = 1, ..., 5, \tag{13.44}$$

$$X \le \overline{G}(p), X \le 3 - M(p), 2 \le M(p) \le 3, X \le \sigma^2(p), 0 \le \sigma^2(p) \le 1. \tag{13.45}$$

All subproblems were solved by SOLVER and the final results, the best one from all four subproblems, is in Table 13.4. More details on SOLVER is in Chapter 30.

The optimal solution for each subproblem is as follows: (1) in subproblem #1 ($2 \le M(p) \le 3, 1 \le \sigma^2(p) \le 2$) $F(p) = 0.900463, M(p) = 2, \overline{A}(p) = 1,$ $\sigma^2(p) = 1.099537, \overline{B}(p) = 0.900463$, $max(X) = 0.900463$; (2) in subproblem #2 ($1 \le M(p) \le 2, 1 \le \sigma^2(p) \le 2$) $F(p) = 0.900463, M(p) = 2, \overline{A}(p) = 1,$ $\sigma^2(p) = 1.099537, \overline{B}(p) = 0.900463$, $max(X) = 0.900463$; (3) in subproblem #3 ($1 \le M(p) \le 2, 0 \le \sigma^2(p) \le 1$) $F(p) = 0.874689, M(p) = 2, \overline{A}(p) = 1,$ $\sigma^2(p) = 1, \overline{B}(p) = 1$, $max(X) = 0.874689$; and (4) in subproblem #4

x	$P(x)$
0	0.0764
1	0.2440
2	0.3592
3	0.2439
4	0.0765

Table 13.4: Solution to Fuzzy Maximum Entropy in Example 13.3.4

$(2 \leq M(p) \leq 3, 0 \leq \sigma^2(p) \leq 1)$ $F(p) = 0.874689$, $M(p) = 2$, $\overline{A}(p) = 1$, $\sigma^2(p) = 1$, $\overline{B}(p) = 1$, $max(X) = 0.874689$. Put these four subproblems together to obtain the overall optimal solution which was the results in both subproblem #1 and #2 whose p_i values are given in Table 13.4.

In comparing to Example 13.3.3 we see that $F(p)$ increased from 0.8747 in Example 13.3.3 to 0.9005 in Example 13.3.4. This was accomplished by relaxing the crisp constraints from $M(p) = 2$ and $\sigma^2(p) = 1$ to $M(p) \approx 2$ and $\sigma^2(p) \approx 1$ with these imprecise constraints modeled by fuzzy numbers.

The above examples may be extended/generalized in various ways. We can easily add more, and different, x_i values. We can also easily add more constraints like (approximate) values for higher order moments of the discrete probability distribution. We do not have to use triangular fuzzy numbers for $\overline{A}$ and $\overline{B}$. For example, let $\overline{A}$ be a triangular shaped fuzzy number with base the interval [2,4] and vertex at $x = 3$, with $y = f_1(x)$ ($y = f_2(x)$) describing the left (right) side of $\overline{A}$. The sides of $\overline{A}$ are now curves, not straight line segments. Then change equation (13.21) in Example 13.3.2 to $X \leq f_1(M(p))$ and equation (13.24) to $X \leq f_2(M(p))$. Surely, we could also employ trapezoidal (shaped) fuzzy numbers.

13.4 Summary and Conclusions

In this chapter we showed how to solve the maximum entropy problem with imprecise side-conditions for a crisp (non-fuzzy) discrete probability distribution. The next step would be to solve for a crisp continuous probability density. That is the topic of the next chapter.

13.5 References

1. R.E. Bellman and L.A. Zadeh: Decision-Making in a Fuzzy Environment, Management Science, 17(1970)B141-B164.

2. J.J. Buckley: Risk Analysis, 5(1985)303-313.

3. J.J. Buckley: Maximum Entropy Principle with Imprecise Side-Conditions II: Crisp Discrete Solutions, Soft Computing. To appear.

4. Frontline Systems (www.frontsys.com).

5. Maple 9, Waterloo Maple Inc., Waterloo, Canada.

6. www.solver.com.

Chapter 14

Max Entropy: Crisp Continuous Solutions

14.1 Introduction

We first discuss the maximum entropy principle, subject to crisp (non-fuzzy) constraints, in the next Section. This presentation is restricted to continuous probability density functions and is based on [2]. This discussion was given in Chapter 12 and repeated here for completeness. Then we show how this principle may be extended to handle fuzzy constraints (fuzzy numbers model the imprecision) in Section 14.3. In Chapter 12 we obtained solutions like fuzzy discrete probability distributions, the fuzzy normal probability distribution, and the fuzzy negative exponential distribution to the maximum entropy problem with fuzzy side-conditions. Then in Chapter 13 we added the constraint that the solution must be non-fuzzy and showed how to solve the problem for discrete probability distributions. This chapter is about the same problem but now we wish to solve it for crisp continuous probability densities. The results are based on [3].

Let $f(x)$ denote a non-fuzzy continuous probability density so that

$$\int_E f(x)dx = 1, f(x) \geq 0, x \in E. \tag{14.1}$$

The discussion will depend on the set E. There will be three cases: (1) $E = [0, M]$, $M > 0$; (2) $E = [0, \infty)$; and (3) $E = (-\infty, \infty)$. The first case is covered in Section 14.4, case (2) in Section 14.5 followed by Section 14.6 devoted to case (3). In each section we first consider the side-condition involving only the mean and then both the mean and the variance. Also, in each section we first solve the crisp problem (non-fuzzy side-condition(s)) and then the fuzzy problem.

James J. Buckley: *Fuzzy Probability and Statistics*, StudFuzz **196**, 125–141 (2006)
www.springerlink.com © Springer-Verlag Berlin Heidelberg 2006

14.2   Max Entropy: Probability Densities

The entropy principle has not gone uncriticized, and this literature, together with that justifying the principles, has been surveyed in [2].

We have a continuous probability density $f(x)$. We do not know an exact formula for $f(x)$ but we do have some prior information, possibly through expert opinion, about the distribution. This information could be in the form of: (1) its mean; and/or (2) its variance. The decision problem is to find the "best" $f(x)$ subject to the constraints given in the information we have about the density. A measure of uncertainty in our decision problem is computed by $\Omega(f)$ where

$$\Omega(f) = -\{\int_E f(x)ln(f(x))dx\}, \tag{14.2}$$

for $f(x) \geq 0$ on E, $\int_E f(x)dx = 1$. Define $0\ln(0) = 0$. $\Omega(f)$ is called the entropy (uncertainty) in the decision problem.

Let $\mathcal{F}$ denote the set of feasible probability density functions f. $\mathcal{F}$ will contain all the f satisfying the constraints dictated by the prior information about the distribution. The maximum entropy principle states that the "best" f, say f^*, has the maximum entropy subject to $f \in \mathcal{F}$. Therefore f^* solves

$$max[-\int_E f(x)ln(f(x))dx], \tag{14.3}$$

subject to $f \in \mathcal{F}$.

Example 14.2.1

Suppose we have prior information, possibly through expert opinion, about the mean m of the density function and its variance σ^2. Our decision problem is

$$max[-\int_E f(x)ln(f(x))dx], \tag{14.4}$$

subject to

$$\int_E f(x) = 1, f(x) \geq 0, x \in E, \tag{14.5}$$

$$\int_E xf(x)dx = m, \tag{14.6}$$

$$\int_E (x - m)^2 f(x)dx = \sigma^2. \tag{14.7}$$

14.3 Max Entropy: Imprecise Side-Conditions

We will only consider imprecise side-conditions relating to the mean and variance of the unknown probability density function. These imprecise conditions will be stated as the mean is "approximately" m and the variance is "approximately" σ^2. We will model this imprecision using triangular fuzzy numbers.

How will we obtain these fuzzy numbers? From expert opinion (Section 3.3)and/or from data (Chapters 6-9). We will use the triangular fuzzy number $\overline{m} = (m_1/m_2/m_3)$ for "approximately" m. Similarly we use $\overline{\sigma}^2 = (\sigma_1^2/\sigma_2^2/\sigma_3^2)$ with $\sigma_1^2 > 0$ for $\approx \sigma^2$.

We now show how to solve the maximum entropy principle with imprecise side-conditions through a series of examples in the next three sections. We first assume that the only prior information is about the mean and we solve both problems: mean=m and mean=$\overline{m}$. Then we assume that the prior information extends to also include the variance and solve both problems. All optimization problems are solved using Maple [5].

14.4 $E = [0, M]$

We start with the crisp problem having no side-conditions. It is $max\ \Omega(f)$ subject to $\int_E f(x)dx = 1$ and $f(x) \geq 0$ for $x \in E$. The solution [2] is the uniform density $f(x) = 1/M$ for $0 \leq x \leq M$. Using the uniform density $\Omega(f) = ln(M)$. Then for any other feasible set $\mathcal{F}$, determined by certain side-conditions, we have $\Omega(f) \leq ln(M)$ for $f \in \mathcal{F}$. We will use this result later on in the section.

Now we add the side-condition involving the mean. The crisp maximum entropy problem is

$$max\ \Omega(f) = -\{\int_0^M f(x)ln(f(x))dx\}, \tag{14.8}$$

subject to

$$\int_0^M f(x)dx = 1, f(x) \geq 0, x \in [0, M], \tag{14.9}$$

and

$$\int_0^M xf(x)dx = m. \tag{14.10}$$

Of course, we assume that m belongs to $(0, M)$. Using the calculus of variations [4] the Euler equation is [2]

$$-ln(f) - 1 + \lambda + \mu x = 0, \tag{14.11}$$

where λ and μ are Lagrange multipliers. The solution is

$$f^*(x) = ce^{\mu x}, \tag{14.12}$$

where $c = e^{\lambda - 1}$. The exact values of λ and μ are obtained from the equations

$$c \int_0^M e^{\mu x} dx = 1, \tag{14.13}$$

$$c \int_0^M x e^{\mu x} dx = m. \tag{14.14}$$

Example 14.4.1

Let $M = 10$ and assume that $m = 3$. Using Maple [5] we solved equations (14.13) and (14.14) for c and μ and obtained $c = 0.28705$ and $\mu = -0.26721$ producing solution

$$f^*(x) = 0.28705 e^{-0.26721x}, 0 \le x \le 10, \tag{14.15}$$

with $max \; \Omega(f) = 2.04974$. We solved this crisp problem to compare to the following fuzzy problem in Example 14.4.2. We will need to normalize $\Omega(f)$ so that its maximum is at most one, for the fuzzy goal for the objective function in the next example, and then $\Omega(f)/ln(10) = 0.89019$ since $M = 10$. The Maple commands for this solution are in Chapter 30.

Now we make the information about the mean imprecise information so that it is

$$\int_0^M x f(x) dx \approx m. \tag{14.16}$$

The prior information was that the mean should be approximately m. We model this imprecise information using a triangular fuzzy number $\overline{A} = (m_1/m/m_2)$. The maximum entropy problem has become

$$max \; \Omega(f) = -\{ \int_0^M f(x) ln(f(x)) dx \}, \tag{14.17}$$

subject to

$$\int_0^M f(x) dx = 1, f(x) \ge 0, x \in [0, M], \tag{14.18}$$

and

$$\int_0^M x f(x) dx = \overline{A}. \tag{14.19}$$

The problem has changed to a fuzzy optimization problem and we will set up fuzzy goals for each objective, then intersect these fuzzy sets, look for the f of maximum membership in the intersection and follow the method given in [1]. We need to construct a fuzzy goal for equation (14.17) and also for equation (14.19).

First consider equation (14.17). The expression $\Omega(f)$ attains its maximum value of $ln(M)$ for $f(x) = 1/M$, $0 \le x \le M$. A fuzzy goal $\overline{G}(f)$ for this

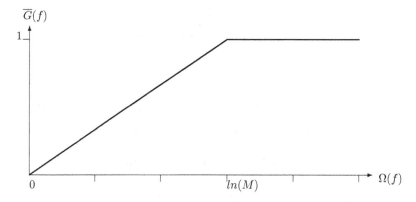

Figure 14.1: Fuzzy Goal $\overline{G}(f)$ in Section 14.4

objective is shown in Figure 14.1. The membership function for this fuzzy goal is $\overline{G}(f) = \Omega(f)/ln(M)$ for $0 \leq \Omega(f) \leq ln(M)$ and $\overline{G}(f) = 1$ if $\Omega(f) \geq ln(M)$.

Next we need to get a fuzzy goal for equation (14.19). Let

$$m = Mean(f) = \int_0^M xf(x)dx, \qquad (14.20)$$

which is the mean of the continuous probability density function $f(x)$. The membership function for the second fuzzy goal is $\overline{A}(f) = \overline{A}(Mean(f)) = \overline{A}(m)$. Then the fuzzy optimization problem becomes

$$max\{min(\overline{G}(f), \overline{A}(f))\}, \qquad (14.21)$$

subject to

$$\int_0^M f(x)dx = 1, f(x) \geq 0, x \in [0, M]. \qquad (14.22)$$

We intersected the two fuzzy goals, denoted by the "min" and then we want the f of maximum membership in $\overline{G}(f) \cap \overline{A}(f)$, subject to the crisp constraints in equation (14.22). The constraints are crisp and there is no need to fuzzify them.

Example 14.4.2

Let $M = 10$ and $\overline{A} = (2/3/4)$. Using Maple [5] we computed values of the fuzzy goal $\overline{G}(f^*)$ for selected values of m in $(0, 5]$. These are shown in Table 14.1. We found that $\overline{G}(f^*)$ was an increasing function of m. Therefore, the $max - min$ solution is easily found and it is shown as point S in Figure 14.2. Figure 14.2 contains the graph of $\overline{A}$ and $\overline{G}(f^*)$ for $2 \leq m \leq 4$. The optimal

m	$\overline{G}(f^*)$
1.0	0.4343
2.0	0.7321
2.5	0.8225
3.0	0.8902
3.4	0.9311
4.0	0.9736
5.0	1.0000

Table 14.1: Values of the Fuzzy Goal $\overline{G}(f^*)$ in Example 14.4.2

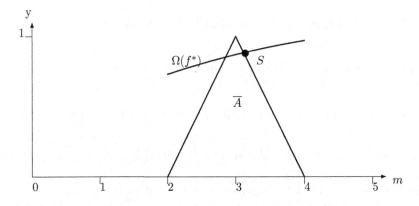

Figure 14.2: Solution to Example 14.4.2

results are $m = 3.09870$, $c = 0.273311$, $\mu = -0.251128$, $\overline{A}(f^*) = \overline{G}(f^*) = 0.90130$ and $f^*(x) = 0.273311e^{-0.251128x}$ for $0 \le x \le 10$.

The next thing to do is add previous information on the variance. The new maximum entropy problem is

$$max\ \Omega(f) = -\{\int_0^M f(x)ln(f(x))dx\}, \tag{14.23}$$

subject to

$$\int_0^M f(x)dx = 1, f(x) \ge 0, x \in [0, M], \tag{14.24}$$

and

$$\int_0^M xf(x)dx = m, \int_0^M (x - m)^2 f(x)dx = s^2. \tag{14.25}$$

Of course, we assume that m belongs to $(0, M)$ and $s^2 > 0$. Using the calculus of variations [4] the Euler equation is [2]

$$-ln(f) - 1 + \lambda + \mu x + \gamma(x - m)^2 = 0, \tag{14.26}$$

where λ, μ and γ are Lagrange multipliers. The solution is

$$f^*(x) = ce^{\mu x}e^{\gamma(x-m)^2}, \tag{14.27}$$

where $c = e^{\lambda - 1}$. The exact values of λ, μ and γ are obtained from the equations

$$c\int_0^M e^{\mu x}e^{\gamma(x-m)^2}dx = 1, \tag{14.28}$$

$$c\int_0^M xe^{\mu x}e^{\gamma(x-m)^2}dx = m, \tag{14.29}$$

$$c\int_0^M (x-m)^2 e^{\mu x}e^{\gamma(x-m)^2} = s^2. \tag{14.30}$$

Example 14.4.3

Let $M = 10$ and assume that $m = 3$ and $s^2 = 1$. Using Maple [5] we solved equations (14.28)-(14.30) for c, μ and γ and obtained $c = 0.40237$, $\mu = -0.00477$ and $\gamma = -0.49285$ producing the optimal solution

$$f^*(x) = 0.40237e^{-0.00477x}e^{-0.49285(x-3)^2}, 0 \le x \le 10, \tag{14.31}$$

with $max \; \Omega(f) = 1.41754$. We solved this crisp problem to compare to the following fuzzy problem in Example 14.4.4. We normalize $\Omega(f)$ so that its maximum is at most one and then $\Omega(f)/ln(10) = 0.61563$ because $M = 10$. Maple commands for this problem are in Chapter 30.

The last thing to do in this section is to make the information about the variance (and the mean) imprecise information so that it is

$$\int_0^M (x-m)^2 f(x)dx \approx s^2. \tag{14.32}$$

The prior information was that the variance should be approximately s^2. We model this imprecise information using a triangular fuzzy number $\overline{B} = (s_1/s^2/s_2)$. The maximum entropy problems has become

$$max \; \Omega(f) = -\{\int_0^M f(x)ln(f(x))dx\}, \tag{14.33}$$

subject to

$$\int_0^M f(x)dx = 1, f(x) \ge 0, x \in [0, M], \tag{14.34}$$

and

$$\int_0^M xf(x)dx = \overline{A}, \int_0^M (x-m)^2 f(x)dx = \overline{B}. \tag{14.35}$$

The problem has changed to a fuzzy optimization problem and we will set up
fuzzy goals for each objective, then intersect these fuzzy sets, look for the f
of maximum membership in the intersection and follow the method given in
[1]. We need to construct a new fuzzy goal for the variance. We will continue
to use the fuzzy goals $\overline{G}(f)$ and $\overline{A}(f)$.

Define

$$s^2 = Var(f) = \int_0^M (x-m)^2 f(x)dx, \qquad (14.36)$$

and then our third fuzzy goal will be $\overline{B}(f) = \overline{B}(Var(f)) = \overline{B}(s^2)$. Then the
fuzzy optimization problem is

$$max\{min(\overline{G}(f), \overline{A}(f), \overline{B}(f))\}, \qquad (14.37)$$

subject to

$$\int_0^M f(x)dx = 1, f(x) \geq 0, x \in [0, M]. \qquad (14.38)$$

We intersected the three fuzzy goals, denoted by the "min" and then we want
the f of maximum membership in $\overline{G}(f) \bigcap \overline{A}(f) \bigcap \overline{B}(f)$, subject to the crisp
constraints. The constraints are crisp and there is no need to fuzzify them.

Finally, we need to consider the triangular fuzzy numbers $\overline{A}$ and $\overline{B}$. We
will do this by making four subproblems and putting the results together to
have the overall solution. We now assume that $\overline{A} = (2/3/4)$ and $\overline{B}(0/3/6)$.
Notice this is a change from before where we had the variance equal to one and
now the variance is approximately three. Looking at the graph of $\overline{A} = (2/3/4)$
we see that the left side of the triangle is the graph of $y = x - 2$ and the right
side is described by $y = 4 - x$. The left side of $\overline{B}$ has the equation $y = \frac{1}{3}x$
and its right side is $y = 2 - \frac{1}{3}x$. Clearly, we need m in the interval $[2, 4]$ and
$s^2 \in [0, 6]$ else the objective function is surely zero.

Partition the rectangle $[2, 4] \times [0, 6]$ up into four subregions: (1) subregion
#1 is $[3, 4] \times [3, 6]$; (2) subregion #2 is $[2, 3] \times [3, 6]$; (3) subregion #3 is
$[2, 3] \times [0, 3]$; and (4) subregion #4 is $[3, 4] \times [0, 3]$. Each subregion corresponds
to a subproblem.

The first subproblem #1 is

$$max\{min(\overline{G}(f), \overline{A}(f), \overline{B}(f))\}, \qquad (14.39)$$

subject to

$$\int_0^M f(x)dx = 1, f(x) \geq 0, x \in [0, M], \qquad (14.40)$$

$$\overline{A}(f) = 4 - m, 3 \leq m \leq 4, \overline{B}(f) = 2 - \frac{1}{3}s^2, 3 \leq s^2 \leq 6. \qquad (14.41)$$

And the second subproblem #2 is

$$max\{min(\overline{G}(f), \overline{A}(f), \overline{B}(f))\}, \qquad (14.42)$$

subject to

$$\int_0^M f(x)dx = 1, f(x) \geq 0, x \in [0, M], \tag{14.43}$$

$$\overline{A}(f) = m - 2, 2 \leq m \leq 3, \overline{B}(f) = 2 - \frac{1}{3}s^2, 3 \leq s^2 \leq 6. \tag{14.44}$$

The third subproblem #3 is

$$max\{min(\overline{G}(f), \overline{A}(f), \overline{B}(f))\}, \tag{14.45}$$

subject to

$$\int_0^M f(x)dx = 1, f(x) \geq 0, x \in [0, M], \tag{14.46}$$

$$\overline{A}(f) = m - 2, 2 \leq m \leq 3, \overline{B}(f) = \frac{1}{3}s^2, 0 \leq s^2 \leq 3. \tag{14.47}$$

And the last subproblem #4 is

$$max\{min(\overline{G}(f), \overline{A}(f), \overline{B}(f))\}, \tag{14.48}$$

subject to

$$\int_0^M f(x)dx = 1, f(x) \geq 0, x \in [0, M], \tag{14.49}$$

$$\overline{A}(f) = 4 - m, 3 \leq m \leq 4, \overline{B}(f) = \frac{1}{3}s^2, 0 \leq s^2 \leq 3. \tag{14.50}$$

We solve for each $m \in [2, 4]$ and each $s^2 \in [0, 6]$ as in Example 14.4.3 producing the solution $f^*(x) = ce^{\mu x}e^{\gamma(x-m)^2}$. Put f^* into $\Omega(f)$ and it simplifies to

$$\Omega(f^*) = -ln(c) - \mu m - \gamma s^2. \tag{14.51}$$

The result is:

1. For each $m \in [3, 4]$, $s^2 \in [3, 6]$ solve for f^* and them compute $max[min([-ln(c) - \mu m - \gamma s^2]/ln(10), 4 - m, 2 - \frac{1}{3}s^2)];.$

2. For each $m \in [2, 3]$, $s^2 \in [3, 6]$, solve for f^* and determine $max[min([-ln(c) - \mu m - \gamma s^2]/ln(10), m - 2, 2 - \frac{1}{3}s^2)].$

3. For each $m \in [2, 3]$, $s^2 \in [0, 3]$, solve for f^* and determine $max[min([-ln(c) - \mu m - \gamma s^2]/ln(10), m - 2, \frac{1}{3}s^2)].$

4. For each $m \in [3, 4]$, $s^2 \in [0, 3]$, solve for f^* and determine $max[min([-ln(c) - \mu m - \gamma s^2]/ln(10), 4 - m, \frac{1}{3}s^2)].$

5. Combine the results into the overall optimal solution.

We illustrate how to solve these problems in the next example.

m	s^2	$\overline{G}(f^*)$	MIN
2.3	1.0	0.6107	0.3
2.3	2.0	0.7301	0.3
2.3	3.0	0.7743	0.3
2.3	4.0	0.7887	0.3
2.3	5.0	0.7862	0.3
2.7	1.0	0.6146	$\frac{1}{3}$
2.7	2.0	0.7498	$\frac{2}{3}$
2.7	3.0	0.8108	0.7
2.7	4.0	0.8402	$\frac{2}{3}$
2.7	5.0	0.8513	$\frac{1}{3}$
3.0	1.0	0.6156	$\frac{1}{3}$
3.0	2.0	0.7574	$\frac{2}{3}$
3.0	3.0	0.8271	0.8271
3.0	4.0	0.8650	$\frac{2}{3}$
3.0	5.0	0.8842	$\frac{1}{3}$
3.3	1.0	0.6160	$\frac{1}{3}$
3.3	2.0	0.7617	$\frac{2}{3}$
3.3	3.0	0.8376	0.7
3.3	4.0	0.8820	$\frac{2}{3}$
3.3	5.0	0.9076	$\frac{1}{3}$
3.7	1.0	0.6162	0.3
3.7	2.0	0.7646	0.3
3.7	3.0	0.8457	0.3
3.7	4.0	0.8963	0.3
3.7	5.0	0.9283	0.3

Table 14.2: Values of the Fuzzy Goals in Example 14.4.4

Example 14.4.4

Let $m = 10$, $\overline{A} = (2/3/4)$ and $\overline{B} = (0/3/6)$. Using Maple [5] we computed values of the fuzzy goal $\overline{G}(f^*)$ for selected values of m and s^2. These are shown in Table 14.2. We must have $s^2 > 0$ and that is why we start its values at 1.0. MIN in Table 14.2 stands for $min(\overline{G}(f^*), \overline{A}(f^*), \overline{B}(f^*))$.

From Table 14.2 we see that: (1) for m fixed in $[2, 4]$ $\overline{G}(f^*)$ is an increasing function of s^2 in $[0, 6]$; and (2) for s^2 fixed in $[0, 6]$ the fuzzy goal is an increasing function of m. It looks like the optimal solution is in region $[3, 4] \times [3, 6]$ and close to the point $m = 3$, $s^2 = 3$. We now computed the fuzzy goal and MIN for values of m and s^2 in this region close to the point $m = 3$, $s^2 = 3$. Our solution is $m = 3.146$, $s^2 = 3.438$, $c = 0.2696$, $\mu = -0.1018$, $\gamma = -0.0974$, $\overline{G}(f^*) = \overline{A}(m) = \overline{B}(s^2) = 0.854$ with

$$f^*(x) = 0.2696e^{-0.1018x}e^{-0.0974(x-3.146)^2}, 0 \leq x \leq 10. \qquad (14.52)$$

Notice that $\overline{G}(f^*)$ increased from 0.6156 in Example 14.4.3 to 0.854 in Example 14.4.4. However we did use the variance approximately 3 in Example 14.4.4 because Maple had trouble solving the equations when s^2 was close to zero. If we used $\overline{B} = (0/1/2)$ for the imprecise variance then we would have s^2 equal to 0.3 with no Maple solution for c, μ and γ.

14.5 $E = [0, \infty)$

We pattern this section after Section 14.4. The only difference is that now $E = [0, \infty)$ so we have to make sure the improper integrals all converge.

First look at the maximum entropy problem with no side-conditions. It is $max\ \Omega(f)$ subject to $\int_0^\infty f(x)dx = 1$ and $f(x) \geq 0$ for $x \geq 0$. The Euler equation is [2]

$$-ln(f) - 1 + \lambda = 0, \tag{14.53}$$

where λ is a Lagrange multiplier. The solution is

$$f^*(x) = c, \tag{14.54}$$

where $c = e^{\lambda-1}$. But the integral of this f^* over $E = [0, \infty)$ surely diverges. The Euler equation has no solution. Also we can show that the objective function $\Omega(f)$ is unbounded ($max = \infty$) over the feasible set $\mathcal{F}$. Clearly the uniform density $f(x) = 1/M$ for $0 \leq x \leq M$ is in $\mathcal{F}$ but for this density $\Omega(f) = ln(M) \to \infty$ as $M \to \infty$. This problem has no solution.

We next consider the crisp problem having one side-condition with respect to its mean. It is $max\ \Omega(f)$ subject to $\int_0^\infty f(x)dx = 1$, $f(x) \geq 0$ for $x \geq 0$ and $\int_0^\infty xf(x)dx = m$. The solution [2] is the exponential density $f^*(x) = (1/m)e^{-x/m}$ for $x \geq 0$. The Euler equation (equation (14.11)), the structure of $f^*(x)$ (equation (14.12)) and the equations (equations (14.13) and (14.14)) to solve for the constants c and μ are all the same except change $E = [0, M]$ to $E = [0, \infty)$. However, we need to assume that $\mu < 0$ so that the improper integrals converge. Using the exponential density $\Omega(f^*) = ln(m) + 1$. Then for any other feasible set $\mathcal{F}$, determined by side-conditions relating to the mean, variance,..., we have $\Omega(f) \leq ln(m) + 1$ for $f \in \mathcal{F}$. We will use this result later on in the section.

Now we add an imprecise side-condition involving the mean. The maximum entropy problem is

$$max\ \Omega(f) = -\{\int_0^\infty f(x)ln(f(x))dx\}, \tag{14.55}$$

subject to

$$\int_0^\infty f(x)dx = 1, f(x) \geq 0, for\ x \geq 0, \tag{14.56}$$

and

$$\int_0^\infty xf(x)dx \approx m. \tag{14.57}$$

Of course, we assume that m belongs to $(0, \infty)$. We model this imprecise information using a triangular fuzzy number $\overline{A} = (m_1/m/m_2)$. The maximum entropy problems has become

$$max \ \Omega(f) = -\{\int_0^\infty f(x)ln(f(x))dx\}, \qquad (14.58)$$

subject to

$$\int_0^\infty f(x)dx = 1, f(x) \geq 0, for \ x \geq 0, \qquad (14.59)$$

and

$$\int_0^\infty xf(x)dx = \overline{A}. \qquad (14.60)$$

The problem has changed to a fuzzy optimization problem and we will set up fuzzy goals for each objective, then intersect these fuzzy sets, look for the f of maximum membership in the intersection and follow the method given in [1]. We use the same fuzzy goals as in Section 14.4. Then the fuzzy optimization problem becomes

$$max\{min(\overline{G}(f), \overline{A}(f))\}, \qquad (14.61)$$

subject to

$$\int_0^\infty f(x)dx = 1, f(x) \geq 0, for \ x \geq 0, \qquad (14.62)$$

We illustrate how to solve the problem in the next example.

Example 14.5.1

Let $\overline{A} = (2/3/4)$. So $2 \leq m \leq 4$, and for each value of m f^* is the exponential, and the maximum $\Omega(f^*)$ can be is $ln(4) + 1$. We normalize $\Omega(f^*)$, so that its maximum will be one, for the fuzzy goal $\overline{G}$ and then $\overline{G}(f) = [ln(m) + 1]/[ln(4) + 1]$. Obviously $\overline{G}(f)$ is an increasing function of m so the optimal solution will be where $y = [ln(m) + 1]/[ln(4) + 1]$ intersects the right side of $\overline{A}$. Hence we must solve

$$[ln(m) + 1]/[ln(4) + 1] = 4 - m, \qquad (14.63)$$

for m in $[3, 4]$. The solution is $m = 3.1060$ producing $f^*(x) = (0.32196)e^{-0.32196x}$, for $x \geq 0$, $\overline{G}(f^*) = 0.8940 = \overline{A}(f^*)$.

The next thing to do is add previous information on the variance. The new maximum entropy problem is

$$max \ \Omega(f) = -\{\int_0^\infty f(x)ln(f(x))dx\}, \qquad (14.64)$$

subject to

$$\int_0^\infty f(x)dx = 1, f(x) \geq 0, for \ x \geq 0, \qquad (14.65)$$

and

$$\int_0^\infty x f(x) dx = m, \int_0^\infty (x-m)^2 f(x) dx = s^2. \tag{14.66}$$

Of course, we assume that $m > 0$ and $s^2 > 0$. The Euler equation (equation (14.26)), the structure of $f^*(x)$ (equation (14.27)) and the equations (equations (14.28)–(14.30)) to solve for the constants c, μ and γ are all the same except change $E = [0, M]$ to $E = [0, \infty)$. However, we need to assume that $\mu < 0$ and $\gamma < 0$ so that the improper integrals converge. Evaluate $\Omega(f)$ at $f = f^*$ and we obtain $\Omega(f^*) = -(\ln(c) + \mu m + \gamma s^2)$.

Example 14.5.2

Assume that $m = 3$ and $s^2 = 1$. Maple [5] solved equations (14.28)–(14.30), using $E = [0, \infty)$, for c, μ and γ and obtained $c = 0.40237$, $\mu = -0.004767$ and $\gamma = -0.49285$ producing the optimal solution

$$f^*(x) = 0.40237 e^{-0.004767x} e^{-0.49285(x-3)^2}, \ x \geq 0, \tag{14.67}$$

with $max\ \Omega(f) = 1.41753$. We solved this crisp problem to compare to the following fuzzy problem in Example 14.5.3. We normalize $\Omega(f)$ so that its maximum is at most one and then $\Omega(f)/[\ln(4) + 1] = 0.59403$, since we previously argued that the maximum was $\ln(m) + 1$. Maple command for this example are in Chapter 30.

The last thing to do in this section is to make the information about the variance (and the mean) imprecise information so that it is

$$\int_0^\infty (x-m)^2 f(x) dx \approx s^2. \tag{14.68}$$

The prior information was that the variance should be approximately s^2. We model this imprecise information using a triangular fuzzy number $\overline{B} = (s_1/s^2/s_2)$. The maximum entropy problems has become

$$max\ \Omega(f) = -\{\int_0^\infty f(x) \ln(f(x)) dx\}, \tag{14.69}$$

subject to

$$\int_0^\infty f(x) dx = 1, f(x) \geq 0, for\ x \geq 0, \tag{14.70}$$

and

$$\int_0^\infty x f(x) dx = \overline{A}, \int_0^\infty (x-m)^2 f(x) dx = \overline{B}. \tag{14.71}$$

We will employ the same fuzzy goals. Then the fuzzy optimization problem becomes

$$max\{min(\overline{G}(f), \overline{A}(f), \overline{B}(f))\}, \tag{14.72}$$

m	s^2	$\overline{G}(f^*)$	MIN
2.3	1.0	0.5893	0.3
2.3	2.0	0.7045	0.3
2.3	3.0	0.7473	0.3
2.3	4.0	0.7634	0.3
2.3	4.8	0.7676	0.3
2.7	1.0	0.5930	$\frac{1}{3}$
2.7	2.0	0.7235	$\frac{2}{3}$
2.7	3.0	0.7824	0.7
2.7	4.0	0.8119	$\frac{2}{3}$
2.7	5.0	0.8265	$\frac{1}{3}$
3.0	1.0	0.5940	$\frac{1}{3}$
3.0	2.0	0.7308	$\frac{2}{3}$
3.0	3.0	0.7982	0.7982
3.0	4.0	0.8356	$\frac{2}{3}$
3.0	5.0	0.8570	$\frac{1}{3}$
3.3	1.0	0.5944	$\frac{1}{3}$
3.3	2.0	0.7350	$\frac{2}{3}$
3.3	3.0	0.8083	0.7
3.3	4.0	0.8519	$\frac{2}{3}$
3.3	5.0	0.8791	$\frac{1}{3}$
3.7	1.0	0.5946	0.3
3.7	2.0	0.7378	0.3
3.7	3.0	0.8162	0.3
3.7	4.0	0.8658	0.3
3.7	5.0	0.8990	0.3

Table 14.3: Values of the Fuzzy Goals in Example 14.5.3

subject to

$$\int_0^\infty f(x)dx = 1, f(x) \geq 0, for\ x \geq 0. \tag{14.73}$$

Example 14.5.3

Let $\overline{A} = (2/3/4)$ and $\overline{B} = (0/3/6)$. Using Maple [5] we computed values of the fuzzy goal $\overline{G}(f^*)$ for selected values of m and s^2. These are shown in Table 14.3. MIN in Table 14.3 stands for $min(\overline{G}(f^*), \overline{A}(f^*), \overline{B}(f^*))$.

From Table 14.3 we see that: (1) for m fixed in $[2, 4]$ $\overline{G}(f^*)$ is an increasing function of s^2 in $[0, 6]$; and (2) for s^2 fixed in $[0, 6]$ the fuzzy goal is an increasing function of m. It looks like the optimal solution is in region $[3, 4] \times [3, 6]$ and close to the point $m = 3$, $s^2 = 3$. We now computed the fuzzy goal and MIN for values of m and s^2 in this region close to the point $m = 3$,

$s^2 = 3$. Our solution is $m = 3.172$, $s^2 = 3.516$, $c = 0.2685$, $\mu = -0.1020$, $\gamma = -0.0962$, $\overline{G}(f^*) = \overline{A}(m) = \overline{B}(s^2) = 0.828$ with

$$f^*(x) = 0.2685e^{-0.1020x}e^{-0.0962(x-3.172)^2}, 0 \le x \le 10. \qquad (14.74)$$

14.6 $E = (-\infty, \infty)$

This section is also patterned after Section 14.4. The difference is that now $E = (-\infty, \infty)$ so we have to make sure all the improper integrals converge.

First look at the maximum entropy problem with no side-conditions. It is max $\Omega(f)$ subject to $\int_{-\infty}^{\infty} f(x)dx = 1$ and $f(x) \ge 0$. As at the beginning of Section 14.5 we may argue that this problem has no solution.

Then we move on to the crisp problem having one side-condition with respect to its mean. It is max $\Omega(f)$ subject to $\int_{-\infty}^{\infty} f(x)dx = 1$, $f(x) \ge 0$ and $\int_{-\infty}^{\infty} xf(x)dx = m$. We now argue that this problem also has no solution. The Euler equation (equation (14.11)), the structure of $f^*(x)$ (equation (14.12)) and the equations (equations (14.13) and (14.14)) to solve for the constants c and μ are all the same except change $E = [0, M]$ to $E = (-\infty, \infty)$. However, equations (14.13) and (14.14) with $E = (-\infty, \infty)$ will never converge. We can not solve for the constants c and μ. But, as we now show, the objective function $\Omega(f)$ is unbounded ($max = \infty$) over the feasible set $\mathcal{F}$. Let $N(m, b)$ denote the normal probability density with mean m and variance b. Clearly $N(m, b) \in \mathcal{F}$. We find that

$$\Omega(N(m, b)) = ln\sqrt{2\pi} + 0.5ln(b) + 0.5. \qquad (14.75)$$

Hence $\Omega(N(m, b)) \to \infty$ as $b \to \infty$. The problem has no solution. Now suppose the side-condition on the mean is imprecise as it should be approximately m. For the reasons given above, it also has no solution.

But the maximum entropy problem with precise/imprecise conditions on both the mean and variance, for $E = (-\infty, \infty)$, does have solutions. First consider

$$max \ \Omega(f) = -\{\int_{-\infty}^{\infty} f(x)ln(f(x))dx\}, \qquad (14.76)$$

subject to

$$\int_{-\infty}^{\infty} f(x)dx = 1, f(x) \ge 0, \qquad (14.77)$$

and

$$\int_{-\infty}^{\infty} xf(x)dx = m, \int_{-\infty}^{\infty} (x - m)^2 f(x)dx = s^2. \qquad (14.78)$$

The solution is $N(m, s^2)$, see [2].

Finally, we make the information about the mean and variance imprecise. The fuzzy optimization problem is

$$max\{min(\overline{G}(f), \overline{A}(f), \overline{B}(f))\}, \qquad (14.79)$$

subject to

$$\int_{-\infty}^{\infty} f(x)dx = 1, f(x) \geq 0. \qquad (14.80)$$

The fuzzy goals $\overline{A}(f)$ and $\overline{B}(f)$ are the same as before just use $E = (-\infty, \infty)$. We need to redefine the fuzzy goal $\overline{G}(f)$.

For any m in the support of $\overline{A}$ and for any s^2 in the support of $\overline{B}$ we know the solution f^* is $N(m, s^2)$. Then $\Omega(f^*) = ln\sqrt{2\pi} + ln(s) + 0.5$, which is independent of m and an increasing function of the variance s^2. Let the support of $\overline{B}$ be the interval (s_1, s_2). Now $s_1 \leq s^2 \leq s_2$ so that $\Omega(f^*) \leq K + ln\sqrt{s_2}$ for $K = ln\sqrt{2\pi} + 0.5$. Hence, we normalize $\Omega(f^*)$ to get $\overline{G}(f)$ and then the membership function for the fuzzy goal is $\overline{G}(f) = [K + ln(s)]/[K + ln\sqrt{s_2}]$ for $s_1 \leq s^2 \leq s_2$ and $\overline{G}(f) = 1$ if $s^2 \geq s_2$.

Example 14.6.1

Let $\overline{A} = (2/3/4)$ and $\overline{B} = (0/1/2)$. Since $\overline{G}(f)$ is independent of m we set $m = 3$ so that $\overline{A}(f) = 1$. Because $\overline{G}(f)$ is an increasing function of s^2 the $max - min$ solution will be where the graph of $\overline{G}(f)$, as a function of s^2, intersects the right side of the graph of $\overline{B}(f)$. Similar to Figure 14.2. We need to solve

$$\overline{G}(f) = 2 - s^2. \qquad (14.81)$$

Using Maple [5] the answer is $s^2 = 1.15540$, $\overline{B}(f) = \overline{G}(f) = 0.8446$. We may take as a solution $N(3, 1.15540)$.

14.7   Summary and Conclusions

We showed how to solve the maximum entropy problem, for a crisp continuous probability density, when the side-conditions, crisp or fuzzy, involved the mean and the variance. We considered three cases for the interval of integration: $[0, M]$, $[0, \infty)$ or $(-\infty, \infty)$. For fuzzy side-conditions we employed the method in [1] to model the optimization problem.

Looking over the numerical examples we can see two major patterns. First, when you change a crisp (non-fuzzy) constraint to a fuzzy constraint the maximum value of the entropy function increases. Secondly, when you add constraints, crisp or fuzzy, the maximum value of the entropy function decreases. The adding of constraints decreases the feasible set.

14.8   References

1. R.E. Bellman and L.A. Zadeh: Decision-Making in a Fuzzy Environment, Management Science, 17(1970)B141-B164.

2. J.J. Buckley: Risk Analysis, 5(1985)303-313.

3. J.J. Buckley: Maximum Entropy Principle with Imprecise Side-Conditions III: Crisp Continuous Solutions, Soft Computing. To appear.

4. I.M. Gelfand and S.V. Fomin: Calculus of Variations, Prentice Hall, Englewood Cliffs, New Jersey, 1963.

5. Maple 9, Waterloo Maple Inc., Waterloo, Canada.

Chapter 15

Tests on μ, Variance Known

15.1 Introduction

This chapter starts a series of chapters (Chapters 15-19,22,25-27) on fuzzy hypothesis testing. In each chapter we first review the crisp case first before proceeding to the fuzzy situation. We give more details on fuzzy hypothesis testing in this chapter.

In the previous chapters on estimation , Chapters 6-11, we sometimes gave multiple graphs of the fuzzy estimators, like for $0.10 \leq \beta \leq 1$ and $0.01 \leq \beta \leq 1$ and $0.001 \leq \beta \leq 1$. We shall not do this again in this book. Unless specified differently, we will always use $0.01 \leq \beta \leq 1$ in the rest of this book. Also, in these previous chapters on estimation we kept reminding the reader about those very short vertical line segments at the two ends of any fuzzy estimator. We will mention this fact only a couple of more times in the rest of this book.

15.2 Non-Fuzzy Case

We obtain a random sample of size n from a $N(\mu, \sigma^2)$, variance σ^2 known, in order to do the following hypothesis test

$$H_0 : \mu = \mu_0, \tag{15.1}$$

verses

$$H_1 : \mu \neq \mu_0. \tag{15.2}$$

In this book we will usually start with the alternate hypothesis H_1 two-sided ($\mu \neq \mu_0$) instead of one-sided ($\mu > \mu_0$ or $\mu < \mu_0$). Using a two-sided alternate hypothesis makes the discussion a little more general, and at the end of this

James J. Buckley: *Fuzzy Probability and Statistics*, StudFuzz **196**, 143–149 (2006)
www.springerlink.com

chapter we will show what changes need to be made for the one-sided tests. From the random sample we compute its mean $\overline{x}$ (a real number), and then determine the statistic

$$z_0 = \frac{\overline{x} - \mu_0}{\sigma/\sqrt{n}}. \tag{15.3}$$

Let γ, $0 < \gamma < 1$, be the significance level of the test. Usual values for γ are $0.10, 0.05, 0.01$. Now under the null hypothesis H_0 z_0 is $N(0, 1)$ (Section 8.2 in [1]) and our decision rule is: (1) reject H_0 if $z_0 \geq z_{\gamma/2}$ or $z_0 \leq -z_{\gamma/2}$; and (2) do not reject H_0 when $-z_{\gamma/2} < z_0 < z_{\gamma/2}$. The numbers $\pm z_{\gamma/2}$ are called the critical values (cv's) for the test. In the above decision rule $z_{\gamma/2}$ is the z-value so that the probability of a random variable, having the $N(0, 1)$ probability density, exceeding z is $\gamma/2$. Usually authors use α for the significance level of a test but in this book we will use α for α-cuts of fuzzy numbers.

15.3   Fuzzy Case

Now proceed to the fuzzy situation where our estimate of μ, as explained in Chapter 6, is the triangular shaped fuzzy number $\overline{\mu}$ where its α-cuts are

$$\overline{\mu}[\alpha] = [\overline{x} - z_{\alpha/2}\sigma/\sqrt{n}, \overline{x} + z_{\alpha/2}\sigma/\sqrt{n}], \tag{15.4}$$

for $0.01 \leq \alpha \leq 1$. In the rest of the book we will always have the base of the fuzzy estimator a 99% confidence interval. Recall that the alpha-cuts of $\overline{\mu}$, for $0 \leq \alpha \leq 0.01$, all equal $\overline{\mu}[0.01]$.

Calculations will be performed by alpha-cuts and interval arithmetic (Sections 2.3.2 and 2.3.3). Our fuzzy statistic becomes

$$\overline{Z} = \frac{\overline{\mu} - \mu_0}{\sigma/\sqrt{n}}. \tag{15.5}$$

Now substitute alpha-cuts of $\overline{\mu}$, equation (15.4), into equation (15.5) and simplify using interval arithmetic producing alpha-cuts of $\overline{Z}$

$$\overline{Z}[\alpha] = [z_0 - z_{\alpha/2}, z_0 + z_{\alpha/2}]. \tag{15.6}$$

We put these α-cuts together to get a triangular shaped fuzzy number $\overline{Z}$.

Since our test statistic is fuzzy the critical values will also be fuzzy. There will be two fuzzy critical value sets: (1) let $\overline{CV}_1$ correspond to $-z_{\gamma/2}$; and (2) let $\overline{CV}_2$ go with $z_{\gamma/2}$. Set $\overline{CV}_i[\alpha] = [cv_{i1}(\alpha), cv_{i2}(\alpha)]$, $i = 1, 2$. We show how to get $cv_{21}(\alpha)$ and $cv_{22}(\alpha)$. The end points of an alpha-cut of $\overline{CV}_2$ are computed from the end points of the corresponding alpha-cut of $\overline{Z}$. We see that to find $cv_{22}(\alpha)$ we solve

$$P(z_0 + z_{\alpha/2} \geq cv_{22}(\alpha)) = \gamma/2, \tag{15.7}$$

for $cv_{22}(\alpha)$. The above equation is the same as

$$P(z_0 \geq cv_{22}(\alpha) - z_{\alpha/2}) = \gamma/2. \tag{15.8}$$

But under H_0 z_0 is $N(0,1)$ so

$$cv_{22}(\alpha) - z_{\alpha/2} = z_{\gamma/2}, \tag{15.9}$$

or

$$c_{22}(\alpha) = z_{\gamma/2} + z_{\alpha/2}. \tag{15.10}$$

By using the left end point of $\overline{Z}[\alpha]$ in equation (15.7) we have

$$cv_{21}(\alpha) = z_{\gamma/2} - z_{\alpha/2}. \tag{15.11}$$

Hence an alpha-cut of $\overline{CV}_2$ is

$$[z_{\gamma/2} - z_{\alpha/2}, z_{\gamma/2} + z_{\alpha/2}]. \tag{15.12}$$

In the above equation for $\overline{CV}_2[\alpha]$, γ is fixed, and α ranges in the interval $[0.01, 1]$. Now $\overline{CV}_1 = -\overline{CV}_2$ so

$$\overline{CV}_1[\alpha] = [-z_{\gamma/2} - z_{\alpha/2}, -z_{\gamma/2} + z_{\alpha/2}]. \tag{15.13}$$

Both $\overline{CV}_1$ and $\overline{CV}_2$ will be triangular shaped fuzzy numbers. When the crisp test statistic has a normal, or t, distribution we will have $\overline{CV}_1 = -\overline{CV}_2$ because these densities are symmetric with respect to zero. However, if the crisp test statistic has the χ^2, or the F, distribution we will have $\overline{CV}_1 \neq -\overline{CV}_2$ because these densities are not symmetric with respect to zero.

Let us present another derivation of $\overline{CV}_2$. Let $0.01 \leq \alpha < 1$ and choose $z \in \overline{Z}[\alpha]$. This value of z is a possible value of the crisp test statistic corresponding to the $(1 - \alpha)100\%$ confidence interval for μ. Then the critical value cv_2 corresponding to z belongs to $\overline{CV}_2[\alpha]$. In fact, as z ranges throughout the interval $\overline{Z}[\alpha]$ its corresponding cv_2 will range throughout the interval $\overline{CV}_2[\alpha]$. Let $z = \tau(z_0 - z_{\alpha/2}) + (1 - \tau)(z_0 + z_{\alpha/2})$ for some τ in $[0, 1]$. Then

$$P(z \geq cv_2) = \gamma/2. \tag{15.14}$$

It follows that

$$P(z_0 \geq cv_2 + (2\tau - 1)z_{\alpha/2}) = \gamma/2. \tag{15.15}$$

Since z_0 is $N(0,1)$ we obtain

$$cv_2 + (2\tau - 1)z_{\alpha/2} = z_{\gamma/2}, \tag{15.16}$$

or

$$cv_2 = \tau(z_{\gamma/2} - z_{\alpha/2}) + (1 - \tau)(z_{\gamma/2} + z_{\alpha/2}), \tag{15.17}$$

which implies that $cv_2 \in \overline{CV}_2[\alpha]$. This same argument can be given in the rest of the book whenever we derive the $\overline{CV}_i$, $i = 1, 2$, however, we shall not go through these details again.

Our final decision (reject, do not reject) will depend on the relationship between $\overline{Z}$ and the $\overline{CV}_i$. Now it is time to review Section 2.5 and Figure 2.5 of Chapter 2. In comparing $\overline{Z}$ and $\overline{CV}_1$ we will obtain $\overline{Z} < \overline{CV}_1$ (reject H_0), or $\overline{Z} \approx \overline{CV}_1$ (no decision), or $\overline{Z} > \overline{CV}_1$ (do not reject). Similar results when comparing $\overline{Z}$ and $\overline{CV}_2$. Let R stand for "reject" H_0, let DNR be "do not reject" H_0 and set ND to be "no decision". After comparing $\overline{Z}$ and the $\overline{CV}_i$ we get (A, B) for $A, B \in \{R, DNR, ND\}$ where A (B) is the result of $\overline{Z}$ verses $\overline{CV}_1$ $(\overline{CV}_2)$. We suggest the final decision to be: (1) if A or B is R, then "reject" H_0; (2) if A and B are both DNR, then "do not reject" H_0, (3) if both A and B are ND (not expected to occur), then we have "no decision"; and (4) if $(A, B) = (ND, DNR)$ or (DNR, ND), then "no decision". The only part of the above decision rule that might be debatable is the fourth one. Users may wish to change the fourth one to "do not reject". However, the author prefers "no decision".

It is interesting that in the fuzzy case we can end up in the "no decision" case. This is because of the fuzzy numbers, which incorporate all the uncertainty in the confidence intervals, that we can get $\overline{M} \approx \overline{N}$ for two different fuzzy numbers $\overline{M}$ and $\overline{N}$.

Let us go through some more details on deciding $\overline{Z} <, \approx, > \overline{CV}_i$ using Section 2.5 and Figure 2.5 before going on to examples. First consider $\overline{Z}$ verses $\overline{CV}_2$. We may have $z_0 > z_{\gamma/2}$ or $z_0 = z_{\gamma/2}$ or $z_0 < z_{\gamma/2}$. First consider $z_0 > z_{\gamma/2}$. So draw $\overline{Z}$ to the right of $\overline{CV}_2$ and find the height of the intersection (the left side of $\overline{Z}$ with the right side of $\overline{CV}_2$). Let the height of the intersection be y_0 as in Section 2.5. If there is no intersection then set $y_0 = 0$. Recall that we are using the test number $\eta = 0.8$ from Section 2.5. The results are: (1) if $y_0 < 0.8$, then $\overline{Z} > \overline{CV}_2$; and (2) if $y_0 \geq 0.8$, then $\overline{Z} \approx \overline{CV}_2$. If $z_0 = z_{\gamma/2}$ then $\overline{Z} \approx \overline{CV}_2$. So now assume that $z_0 < z_{\gamma/2}$. The height of the intersection y_0 will be as shown in Figure 2.5. The decision is: (1) if $y_0 < 0.8$, then $\overline{Z} < \overline{CV}_2$; and (2) if $y_0 \geq 0.8$, then $\overline{Z} \approx \overline{CV}_2$. Similar results hold for $\overline{Z}$ verses $\overline{CV}_1$.

A summary of the cases we expect to happen are: (1) $\overline{CV}_2 < \overline{Z}$ reject; (1) $\overline{CV}_1 < \overline{Z} \approx \overline{CV}_2$ no decision; (3) $\overline{CV}_1 < \overline{Z} < \overline{CV}_2$ do not reject; (4) $\overline{CV}_1 \approx \overline{Z} < \overline{CV}_2$ no decision; and (5) $\overline{Z} < \overline{CV}_1$ reject.

Example 15.3.1

Assume that $n = 100$, $\mu_0 = 1$, $\sigma = 2$ and the significance level of the test is $\gamma = 0.05$ so $z_{\gamma/2} = 1.96$. From the random sample let $\overline{x} = 1.32$ and we then compute $z_0 = 1.60$. Recall that we will be using $0.01 \leq \alpha \leq 1$ in these chapters on fuzzy hypothesis testing.

Since $z_0 < z_{\gamma/2}$ we need to compare the right side of $\overline{Z}$ to the left side of

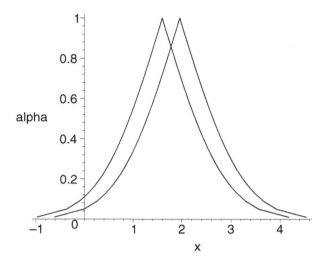

Figure 15.1: Fuzzy Test $\overline{Z}$ verses $\overline{CV}_2$ in Example 15.3.1($\overline{Z}$ left, $\overline{CV}_2$ right)

$\overline{CV}_2$. These are shown in Figure 15.1. The Maple [2] commands for Figure 15.1 are in Chapter 30. The right side of $\overline{Z}$ decreases form $(1.60, 1)$ towards zero and the left side of $\overline{CV}_2$ increases from just above zero to $(1.96, 1)$. Draw a horizontal line through 0.8 on the vertical axis and we observe that the height of intersection is greater than 0.8. From this comparison we conclude $\overline{Z} \approx \overline{CV}_2$.

Next we evaluate $\overline{Z}$ verses $\overline{CV}_1$. The results are in Figure 15.2. Since the height of the intersection is less than 0.8 we conclude $\overline{CV}_1 < \overline{Z}$. Our final conclusion is no decision on H_0. In the crisp case since $-z_{\gamma/2} < z_0 < z_{\gamma/2}$ we would decide: do not reject H_0.

Example 15.3.2

All the data is the same as in Example 15.3.1 except now assume that $\overline{x} = 0.40$. We compute $z_0 = -3.0$. Since $z_0 < -z_{\gamma/2}$ we first compare $\overline{Z}$ and $\overline{CV}_1$. This is shown in Figure 15.3. The right side of $\overline{Z}$ decreases from $(-3.0, 1)$ towards zero and the left side of $\overline{CV}_1$ increases from near zero to $(-1.96, 1)$. The height of the intersection is clearly less than 0.8 so $\overline{Z} < \overline{CV}_1$. It is obvious that we also have $\overline{Z} < \overline{CV}_2$. Hence we reject H_0. In the crisp case we would also reject H_0.

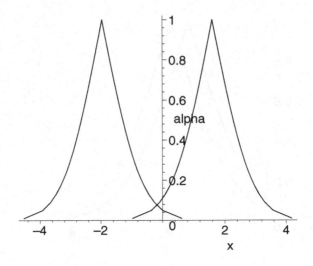

Figure 15.2: Fuzzy Test $\overline{Z}$ verses $\overline{CV}_1$ in Example 15.3.1($\overline{CV}_1$ left, $\overline{Z}$ right)

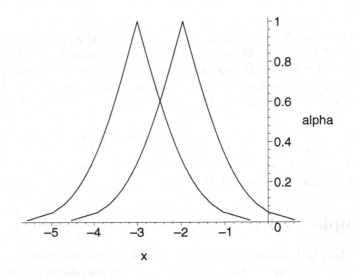

Figure 15.3: Fuzzy Test $\overline{Z}$ verses $\overline{CV}_1$ in Example 15.3.2($\overline{Z}$ left, $\overline{CV}_1$ right)

15.4 One-Sided Tests

First consider

$$H_0 : \mu = \mu_0, \tag{15.18}$$

verses

$$H_1 : \mu > \mu_0. \tag{15.19}$$

Then we would reject H_0 when $z_0 \geq z_\gamma$, here we use γ and not $\gamma/2$, and do not reject if $z_0 < z_\gamma$.

In the fuzzy case we still have the same fuzzy statistic $\overline{Z}$, no $\overline{CV}_1$, and now $\overline{CV}_2$ would be centered (membership value one) at z_γ. The final decision would be made only through the comparison of $\overline{Z}$ and $\overline{CV}_2$. The three case are: (1) if $\overline{Z} < \overline{CV}_2$, then do not reject, (2) if $\overline{Z} \approx \overline{CV}_2$, then no decision on H_0; and (3) if $\overline{Z} > \overline{CV}_2$, then reject H_0.

Obvious changes are to be made if we have the other one-sided test

$$H_0 : \mu = \mu_0, \tag{15.20}$$

verses

$$H_1 : \mu < \mu_0. \tag{15.21}$$

We shall use these one-sided fuzzy hypothesis tests in Chapters 22 and 25.

15.5 References

1. R.V. Hogg and E.A. Tanis: Probability and Statistical Inference, Sixth Edition, Prentice Hall, Upper Saddle River, N.J., 2001.

2. Maple 9, Waterloo Maple Inc., Waterloo, Canada.

Chapter 16

Tests on μ, Variance Unknown

16.1 Introduction

We start with a review of the crisp case and then present the fuzzy model.

16.2 Crisp Case

We obtain a random sample of size n from a $N(\mu, \sigma^2)$, mean and variance unknown, in order to do the following hypothesis test

$$H_0 : \mu = \mu_0, \tag{16.1}$$

verses

$$H_1 : \mu \neq \mu_0. \tag{16.2}$$

From the random sample we compute its mean $\overline{x}$, this will be a real number and not a fuzzy set, and the sample variance s^2 and then determine the statistic

$$t_0 = \frac{\overline{x} - \mu_0}{s/\sqrt{n}}. \tag{16.3}$$

Let γ, $0 < \gamma < 1$, be the significance level of the test. Under the null hypothesis H_0 t_0 has a t-distribution with $n-1$ degrees of freedom (Section 8.2 of [1]) and our decision rule is: (1) reject H_0 if $t_0 \geq t_{\gamma/2}$ or $t_0 \leq -t_{\gamma/2}$; and (2) do not reject H_0 when $-t_{\gamma/2} < t_0 < t_{\gamma/2}$. The numbers $\pm t_{\gamma/2}$ are called the critical values (cv's) for the test. In the above decision rule $t_{\gamma/2}$ is the t-value so that the probability of a random variable, having a t-distribution with $n-1$ degrees of freedom, exceeding t is $\gamma/2$.

James J. Buckley: *Fuzzy Probability and Statistics*, StudFuzz **196**, 151–157 (2006)
www.springerlink.com   © Springer-Verlag Berlin Heidelberg 2006

16.3   Fuzzy Model

Now proceed to the fuzzy situation where our statistic for t_0 will become fuzzy $\overline{T}$. In equation (16.3) for t_0 substitute an alpha-cut of our fuzzy estimator $\overline{\mu}$ for $\overline{x}$, equation (7.4) of Chapter 7; and substitute an alpha-cut of our fuzzy estimator $\overline{\sigma}$ for s, the square root of equation (9.10) of Chapter 9. Use interval arithmetic to simplify and we obtain

$$\overline{T}[\alpha] = [\Pi_1(t_0 - t_{\alpha/2}), \Pi_2(t_0 + t_{\alpha/2})], \tag{16.4}$$

where

$$\Pi_1 = \sqrt{\frac{R(\lambda)}{n-1}}, \tag{16.5}$$

and

$$\Pi_2 = \sqrt{\frac{L(\lambda)}{n-1}}, \tag{16.6}$$

for $L(\lambda)$ $(R(\lambda))$ defined in equation (9.8) $((9.9))$ of Chapter 9. In the interval arithmetic employed above, we assumed that all intervals were positive ($[a, b] > 0$ if $a > 0$). It may happen that for certain values of alpha: (1) the interval in the numerator the left end point is negative, the other end point is positive, but the other interval in the denominator is positive; or (2) the interval in the numerator the right end point of the interval is negative, and so the other end point is also negative, and the other interval in the denominator is positive. These special cases will be discussed at the end of this section. For now we assume that all intervals are positive.

Since our test statistic is fuzzy the critical values will also be fuzzy. There will be two fuzzy critical value sets: (1) let $\overline{CV}_1$ correspond to $-t_{\gamma/2}$; and (2) let $\overline{CV}_2$ go with $t_{\gamma/2}$. Set $\overline{CV}_i[\alpha] = [cv_{i1}(\alpha), cv_{i2}(\alpha)]$, $i = 1, 2$. We show how to get $cv_{21}(\alpha)$ and $cv_{22}(\alpha)$. The end points of an alpha-cut of $\overline{CV}_2$ are computed from the end points of the corresponding alpha-cut of $\overline{T}$. We see that to find $cv_{22}(\alpha)$ we solve

$$P(\Pi_2(t_0 + t_{\alpha/2}) \geq cv_{22}(\alpha)) = \gamma/2, \tag{16.7}$$

for $cv_{22}(\alpha)$. The above equation is the same as

$$P(t_0 \geq (cv_{22}(\alpha)/\Pi_2) - t_{\alpha/2}) = \gamma/2. \tag{16.8}$$

But t_0 has a t distribution so

$$(cv_{22}(\alpha)/\Pi_2) - t_{\alpha/2} = t_{\gamma/2}, \tag{16.9}$$

or

$$c_{22}(\alpha) = \Pi_2(t_{\gamma/2} + t_{\alpha/2}). \tag{16.10}$$

By using the left end point of $\overline{T}[\alpha]$ we have

$$cv_{21}(\alpha) = \Pi_1(t_{\gamma/2} - t_{\alpha/2}). \tag{16.11}$$

Hence an alpha-cut of $\overline{CV}_2$ is

$$[\Pi_1(t_{\gamma/2} - t_{\alpha/2}), \Pi_2(t_{\gamma/2} + t_{\alpha/2})]. \tag{16.12}$$

In the above equation for $\overline{CV}_2[\alpha]$, γ is fixed, and α ranges in the interval $[0.01, 1]$. Now $\overline{CV}_1 = -\overline{CV}_2$ so

$$\overline{CV}_1[\alpha] = [\Pi_2(-t_{\gamma/2} - t_{\alpha/2}), \Pi_1(-t_{\gamma/2} + t_{\alpha/2})]. \tag{16.13}$$

Both $\overline{CV}_1$ and $\overline{CV}_2$ will be triangular shaped fuzzy numbers.

The details of comparing $\overline{T}$ to $\overline{CV}_1$ and $\overline{CV}_2$, and our method of coming to a final decision (reject, no decision, or do not reject), is all in Chapter 15. Before we can go on to work two examples we need to solve a major problem with these fuzzy numbers $\overline{T}$ and the $\overline{CV}_i$: their α-cuts depend on two variables λ and α. $R(\lambda)$ and $L(\lambda)$ are obviously functions of λ and $t_{\alpha/2}$ is a function of α. But, as pointed out in Chapter 9 $\alpha = f(\lambda)$, or α is a function of λ. This comes from equations (9.12) and (9.13) of Chapter 9

$$\alpha = f(\lambda) = \int_0^{R(\lambda)} \chi^2 dx + \int_{L(\lambda)}^\infty \chi^2 dx, \tag{16.14}$$

as λ goes from zero to one. The chi-square density in the integrals has $n - 1$ degrees of freedom. When $\lambda = 0$ then $\alpha = 0.01$ and if $\lambda = 1$ so does $\alpha = 1$. Notice that $\overline{T}[1] = [t_0, t_0] = t_0$, $\overline{CV}_1[1] = [-t_{\gamma/2}, -t_{\gamma/2}] = -t_{\gamma/2}$ and $\overline{CV}_2[1] = [t_{\gamma/2}, t_{\gamma/2}] = t_{\gamma/2}$.

To generate the triangular shaped fuzzy number $\overline{T}$ from its α-cuts in equation (16.4) we increase λ from zero to one (this determines the Π_i), compute α from λ in equation (16.14), which gives $t_{\alpha/2}$, and we have the alpha-cuts since t_0 is a constant. Put these alpha-cuts together to have $\overline{T}$. Similar remarks for the other two triangular shaped fuzzy numbers $\overline{CV}_i$, $i = 1, 2$.

16.3.1 $\overline{T}[\alpha]$ for Non-Positive Intervals

In finding equation (16.4) we divided two intervals

$$\overline{T}[\alpha] = \frac{[a, b]}{[c, d]}, \tag{16.15}$$

where

$$a = \overline{x} - t_{\alpha/2}s/\sqrt{n} - \mu_0, \tag{16.16}$$

and

$$b = \overline{x} + t_{\alpha/2}s/\sqrt{n} - \mu_0, \tag{16.17}$$

and

$$c = \sqrt{\frac{n-1}{L(\lambda)}}(s/\sqrt{n}), \tag{16.18}$$

and

$$d = \sqrt{\frac{n-1}{R(\lambda)}}(s/\sqrt{n}). \qquad (16.19)$$

Now c and d are always positive but a or b could be negative for certain values of alpha.

First assume that $t_0 > 0$ so that $b > 0$ also. Also assume that there is a value of alpha, say α^* in $(0,1)$, so that $a < 0$ for $0.01 \le \alpha < \alpha^*$ but $a > 0$ for $\alpha^* < \alpha \le 1$. Then we compute, using interval arithmetic,

$$\overline{T}[\alpha] = [a,b][1/d, 1/c] = [a/c, b/c], \qquad (16.20)$$

for $0.01 \le \alpha < \alpha^*$ when $a < 0$ and

$$\overline{T}[\alpha] = [a,b][1/d, 1/c] = [a/d, b/c], \qquad (16.21)$$

when $\alpha^* < \alpha \le 1$ for $a > 0$. This case will be in the following Example 16.3.1. The $a > 0$ was what we were using above for equation (16.4). We saw in Chapter 15 that the alpha-cuts of the fuzzy statistic determines the alpha-cuts of the fuzzy critical values. In this case $\overline{T}[\alpha]$ will determine $\overline{CV}_2[\alpha]$. So if we change how we compute $\overline{T}[\alpha]$ when $a < 0$, then we need to use this to find the new α-cuts of $\overline{CV}_2$ and then $\overline{CV}_1 = -\overline{CV}_2$.

Now if $t_0 < 0$ then $a < 0$ but b may be positive for some alpha and negative for other α. So assume that $b < 0$ for $0 < \alpha^* < \alpha \le 1$ and $b > 0$ otherwise. then

$$\overline{T}[\alpha] = [a,b][1/d, 1/c] = [a/c, b/d], \qquad (16.22)$$

when $b < 0$ and

$$\overline{T}[\alpha] = [a,b][1/d, 1/c] = [a/c, b/c], \qquad (16.23)$$

for $b > 0$. In Example 16.3.2 below we will be interested in the $b < 0$ case (because $t_0 < 0$). The alpha-cuts of the test statistic will determine those of the fuzzy critical values. In this case $\overline{T}[\alpha]$ determines $\overline{CV}_1[\alpha]$. When we change how we get alpha-cuts of $\overline{T}$ when $b > 0$ we use this to compute the new α-cuts of $\overline{CV}_1$ and $\overline{CV}_2 = -\overline{CV}_1$.

Assuming $a < 0$ and $b < 0$ (at least for $0 < \alpha^* < \alpha \le 1$) since $t_0 < 0$, then

$$\overline{T}[\alpha] = [\Pi_2(t_0 - t_{\alpha/2}), \Pi_1(t_0 + t_{\alpha/2})], \qquad (16.24)$$

and

$$\overline{CV}_1[\alpha] = [\Pi_2(-t_{\gamma/2} - t_{\alpha/2}), \Pi_1(-t_{\gamma/2} + t_{\alpha/2})], \qquad (16.25)$$

and $\overline{CV}_2 = -\overline{CV}_1$ so that

$$\overline{CV}_2[\alpha] = [\Pi_1(t_{\gamma/2} - t_{\alpha/2}), \Pi_2(t_{\gamma/2} + t_{\alpha/2})], \qquad (16.26)$$

These equations will be used in the second example below.

Example 16.3.1

Assume that the random sample size is $n = 101$ and $\mu_0 = 1$, $\gamma = 0.01$ so that $t_{\gamma/2} = 2.626$ with 100 degrees of freedom. From the data suppose we found $\overline{x} = 1.32$ and the sample variance is $s^2 = 4.04$. From this we compute $t_0 = 1.60$. In order to find the Π_i we obtain $\chi^2_{R,0.005} = 140.169$ and $\chi^2_{L,0.005} = 67.328$. Then $L(\lambda) = 140.169 - 40.169\lambda$ and $R(\lambda) = 67.328 + 32.672\lambda$ and

$$\Pi_2 = \sqrt{1.40169 - 0.40169\lambda}, \tag{16.27}$$

and

$$\Pi_1 = \sqrt{0.67329 + 0.32672\lambda}. \tag{16.28}$$

Using $\alpha = f(\lambda)$ from equation (16.14) we may now have Maple [2] do the graphs of $\overline{T}$ and the $\overline{CV}_i$. Notice that in equations (16.4),(16.12) and (16.13) for the alpha-cuts instead of $t_{\alpha/2}$ we use $t_{f(\lambda)/2}$.

In comparing $\overline{T}$ and $\overline{CV}_2$ we can see the result in Figure 16.1. The Maple commands for this figure are in Chapter 30. Since $t_0 < t_{\gamma/2}$ we only need to compare the right side of $\overline{T}$ to the left side of $\overline{CV}_2$. The height of the intersection is $y_0 > 0.8$ (the point of intersection is close to, by just greater than 0.8)and we conclude that $\overline{T} \approx \overline{CV}_2$ with no decision.

The graphs in Figure 16.1 are correct only to the right of the vertical axis. The left side of $\overline{T}$ goes negative, and we should make the adjustment described in Subsection 16.3.1 to both $\overline{T}$ and $\overline{CV}_2$, but we did not because it will not effect our conclusion.

Next we compare $\overline{T}$ to $\overline{CV}_1$. We need to take into consideration here that the left side of $\overline{T}$ will go negative (see subsection on this topic just preceding

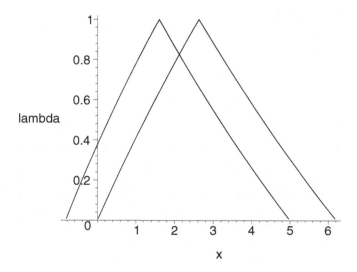

Figure 16.1: Fuzzy Test $\overline{T}$ verses $\overline{CV}_2$ in Example 16.3.1($\overline{T}$ left, $\overline{CV}_2$ right)

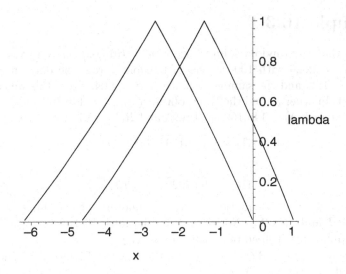

Figure 16.2: Fuzzy Test $\overline{T}$ verses $\overline{CV}_1$ in Example 16.3.2($\overline{T}$ right, $\overline{CV}_1$ left)

this example). Noting this fact we find that the height of the intersection (the left side of $\overline{T}$ with the right side of $\overline{CV}_1$) is less than 0.8. Hence $\overline{CV}_1 < \overline{T}$ with do not reject. It is easy to see the relationship between $\overline{T}$ and $\overline{CV}_1$ because $\overline{CV}_1 = -\overline{CV}_2$.

Our final conclusion is no decision on H_0. In the crisp case the conclusion would be do not reject since $-t_{\gamma/2} < t_0 < t_{\gamma/2}$.

Example 16.3.2

Assume the data is the same as in Example 16.3.1 except $\overline{x} = 0.74$ so that $t_0 = -1.3$. Now we use the results for $a < 0$ and $b < 0$ (for $0 < \alpha^* < \alpha \leq 1$) discussed in the Subsection 16.3.1 above. Since $-t_{\gamma/2} < t_0 < 0$ we first compare the left side of $\overline{T}$ to the right side of $\overline{CV}_1$. We see that the height of the intersection is close to, but less than, 0.8. Hence $\overline{CV}_1 < \overline{T}$ and we do not reject H_0. Again, the graph of $\overline{T}$ in Figure 16.2 is not accurate to the right of the vertical axis, which also effects the right side of $\overline{CV}_1$, but this does not change the conclusion.

Next we compare the right side of $\overline{T}$, taking into account that now b will become positive, with the left side of $\overline{CV}_2$. We determine that $\overline{T} < \overline{CV}_2$ since $\overline{CV}_2 = -\overline{CV}_1$ and we do not reject H_0.

Out final conclusion is to not reject H_0. In the crisp case we also do not reject.

16.4 References

1. R.V. Hogg and E.A. Tanis: Probability and Statistical Inference, Sixth Edition, Prentice Hall, Upper Saddle River, N.J., 2001.

2. Maple 9, Waterloo Maple Inc., Waterloo, Canada.

Chapter 17

Tests on p for a Binomial Population

17.1 Introduction

We begin with the non-fuzzy test and then proceed to the fuzzy test.

17.2 Non-Fuzzy Test

We obtain a random sample of size n from a binomial in order to perform the following hypothesis test

$$H_0 : p = p_0, \tag{17.1}$$

verses

$$H_1 : p \neq p_0. \tag{17.2}$$

From the random sample we compute an estimate of p. In the binomial p represents the probability of a "success". Suppose we obtained x successes in n trials so $\widehat{p} = x/n$ is our point estimate of p. Then determine the statistic

$$z_0 = \frac{\widehat{p} - p_0}{\sqrt{p_0 q_0 / n}}, \tag{17.3}$$

where $q_0 = 1 - p_0$. If n is sufficiently large, say $n > 30$, then under the null hypothesis we may use the normal approximation to the binomial and z_0 is approximately $N(0, 1)$ (Section 8.1 of [1]).

Let γ, $0 < \gamma < 1$, be the significance level of the test. Our decision rule is: (1) reject H_0 if $z_0 \geq z_{\gamma/2}$ or $z_0 \leq -z_{\gamma/2}$; and (2) do not reject H_0 when $-z_{\gamma/2} < z_0 < z_{\gamma/2}$.

James J. Buckley: *Fuzzy Probability and Statistics*, StudFuzz **196**, 159–162 (2006)
www.springerlink.com   © Springer-Verlag Berlin Heidelberg 2006

17.3   Fuzzy Test

In Chapter 8 our fuzzy estimator for p is a triangular shaped fuzzy number $\overline{p}$ where its α-cuts are

$$\overline{p}[\alpha] = [\widehat{p} - z_{\alpha/2}\sqrt{\widehat{pq}/n}, \widehat{p} + z_{\alpha/2}\sqrt{\widehat{pq}/n}], \tag{17.4}$$

where $\widehat{q} = 1 - \widehat{p}$.

Calculations will be performed by alpha-cuts and interval arithmetic. Substitute $\overline{p}$ for $\widehat{p}$ in equation (17.3) and we obtain the following alpha-cut of our fuzzy statistic $\overline{Z}$

$$\overline{Z}[\alpha] = [z_0 - z_{\alpha/2}\sqrt{\frac{\widehat{pq}}{p_0 q_0}}, z_0 + z_{\alpha/2}\sqrt{\frac{\widehat{pq}}{p_0 q_0}}]. \tag{17.5}$$

The critical region will now be determined by fuzzy critical values $\overline{CV}_i$, $i = 1, 2$. They are determined as in the previous two chapters and they are given by their alpha-cuts

$$\overline{CV}_2[\alpha] = [z_{\gamma/2} - z_{\alpha/2}\sqrt{\frac{\widehat{pq}}{p_0 q_0}}, z_{\gamma/2} + z_{\alpha/2}\sqrt{\frac{\widehat{pq}}{p_0 q_0}}], \tag{17.6}$$

all α where γ is fixed, and because $\overline{CV}_1 = -\overline{CV}_2$

$$\overline{CV}_1[\alpha] = [-z_{\gamma/2} - z_{\alpha/2}\sqrt{\frac{\widehat{pq}}{p_0 q_0}}, -z_{\gamma/2} + z_{\alpha/2}\sqrt{\frac{\widehat{pq}}{p_0 q_0}}]. \tag{17.7}$$

Now that we have $\overline{Z}$, $\overline{CV}_1$ and $\overline{CV}_2$ we may compare $\overline{Z}$ and $\overline{CV}_1$ and then compare $\overline{Z}$ with $\overline{CV}_2$. The final decision rule was presented in Chapter 15.

Example 17.3.1

Let $n = 100$, $p_0 = q_0 = 0.5$, $\widehat{p} = 0.54$ so $\widehat{q} = 0.46$, and $\gamma = 0.05$. We then calculate $z_{\gamma/2} = 1.96$, and $z_0 = 0.80$. This is enough information to construct the fuzzy numbers $\overline{Z}$, $\overline{CV}_1$ and $\overline{CV}_2$ from Maple [2].

We first compare $\overline{Z}$ and $\overline{CV}_2$ to see which $\overline{Z} < \overline{CV}_2$, $\overline{Z} \approx \overline{CV}_2$ or $\overline{Z} > \overline{CV}_2$ is true. Since $z_0 < z_{\gamma/2}$ all we need to do is compare the right side of $\overline{Z}$ to the left side of $\overline{CV}_2$. From Figure 17.1 we see that the height of the intersection is less than 0.8 so we conclude that $\overline{Z} < \overline{CV}_2$. The Maple commands for Figure 17.1 are in Chapter 30.

Next we compare $\overline{Z}$ and $\overline{CV}_1$. Since $z_0 > -z_{\gamma/2}$ we compare the left side of $\overline{Z}$ to the right side of $\overline{CV}_1$. We easily see that $\overline{CV}_1 < \overline{Z}$.

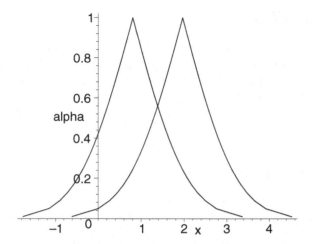

Figure 17.1: Fuzzy Test $\overline{Z}$ verses $\overline{CV}_2$ in Example 17.3.1($\overline{Z}$ left, $\overline{CV}_2$ right)

We do not reject H_0 because $\overline{CV}_1 < \overline{Z} < \overline{CV}_2$, which is the same result as in the crisp test.

We note again, as in previous chapters, that the graph of $\overline{Z}$ in Figure 17.1 is not accurate to the left of the vertical axis. This will also effect the left side of $\overline{CV}_2$. However, this does not effect the conclusion.

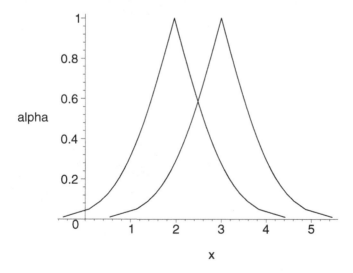

Figure 17.2: Fuzzy Test $\overline{Z}$ verses $\overline{CV}_2$ in Example 17.3.2($\overline{Z}$ right, $\overline{CV}_2$ left)

Example 17.3.2

Assume all the data is the same as in Example 17.3.1 except $\widehat{p} = 0.65$ so $\widehat{q} = 0.35$. We compute $z_0 = 3.00$. Now $z_0 > z_{\gamma/2} = 1.96$ so let us start with comparing the left side of $\overline{Z}$ to the right side of $\overline{CV}_2$. This is shown in Figure 17.2. We decide that $\overline{CV}_2 < \overline{Z}$ because the height of the intersection in Figure 17.2 is less than 0.8. Clearly we also obtain $\overline{CV}_1 < \overline{Z}$. Hence, we reject H_0.

17.4   References

1. R.V. Hogg and E.A. Tanis: Probability and Statistical Inference, Sixth Edition, Prentice Hall, Upper Saddle River, N.J., 2001.

2. Maple 9, Waterloo Maple Inc., Waterloo, Canada.

Chapter 18

Tests on σ^2, Normal Population

18.1 Introduction

We first describe the crisp hypothesis test and then construct the fuzzy test.

18.2 Crisp Hypothesis Test

We obtain a random sample of size n from a $N(\mu, \sigma^2)$, variance σ^2 unknown, in order to do the following hypothesis test

$$H_0 : \sigma^2 = \sigma_0^2, \tag{18.1}$$

verses

$$H_1 : \sigma^2 \neq \sigma_0^2. \tag{18.2}$$

From the random sample we compute its sample variance s^2 and then determine the statistic

$$\chi_0^2 = \frac{(n-1)s^2}{\sigma_0^2}. \tag{18.3}$$

Let γ, $0 < \gamma < 1$, be the significance level of the test. Now under the null hypothesis H_0 χ_0^2 has a chi-square distribution with $(n-1)$ degrees of freedom (Section 8.2 in [1]). Our decision rule is: (1) reject H_0 if $\chi_0^2 \geq \chi_{R,\gamma/2}^2$ or $\chi^2 \leq \chi_{L,\gamma/2}^2$; and (2) do not reject H_0 when $\chi_{L,\gamma/2}^2 < \chi_0^2 < \chi_{R,\gamma/2}^2$. In the above decision rule $\chi_{L,\gamma/2}^2$ $(\chi_{R,\gamma/2}^2)$ is the χ^2-value, on the left (right) side of the density, so that the probability of less than (exceeding) it is $\gamma/2$.

James J. Buckley: *Fuzzy Probability and Statistics*, StudFuzz **196**, 163–166 (2006)
www.springerlink.com

18.3   Fuzzy Hypothesis Test

Our fuzzy estimator of σ^2 is the triangular shaped fuzzy number $\overline{\sigma}^2$ having α-cuts given by equation (9.10) in Chapter 9. Now substitute $\overline{\sigma}^2$ in for s^2 in equation (18.3) and simplify using interval arithmetic and we obtain our fuzzy statistic $\overline{\chi}^2$ whose alpha-cuts are given by

$$\overline{\chi}^2[\alpha] = [\frac{n-1}{L(\lambda)}\chi_0^2, \frac{n-1}{R(\lambda)}\chi_0^2], \tag{18.4}$$

where $L(\lambda)$ $(R(\lambda))$ was defined in equation (9.8) ((9.9)) in Chapter 9. Recall that α is determined from λ, $0 \leq \lambda \leq 1$, as shown in equations (9.12) and (9.13) in Chapter 9.

Using this fuzzy statistic we determine the two fuzzy critical value sets whose alpha-cuts are

$$\overline{CV}_1[\alpha] = [\frac{n-1}{L(\lambda)}\chi_{L,\gamma/2}^2, \frac{n-1}{R(\lambda)}\chi_{L,\gamma/2}^2], \tag{18.5}$$

and

$$\overline{CV}_2[\alpha] = [\frac{n-1}{L(\lambda)}\chi_{R,\gamma/2}^2, \frac{n-1}{R(\lambda)}\chi_{R,\gamma/2}^2]. \tag{18.6}$$

In the above equations for the $\overline{CV}_i$ γ is fixed and λ ranges from zero to one. Notice that now $\overline{CV}_1 \neq -\overline{CV}_2$. When using the normal, or the t, distribution we were able to use $\overline{CV}_1 = -\overline{CV}_2$ because those densities were symmetric with respect to zero.

Having constructed these fuzzy numbers we go on to deciding on $\overline{\chi}^2 < \overline{CV}_1, \dots, \overline{\chi}^2 > \overline{CV}_2$ and then our method of making the final decision (reject, do not reject, no decision) was outlined in Chapter 15.

Example 18.3.1

Let $n = 101$, $\sigma_0^2 = 2$, $\gamma = 0.01$ and from the random sample $s^2 = 1.675$. Then compute $\chi_0^2 = 83.75$, $\chi_{L,0.005}^2 = 67.328$, $\chi_{R,0.005}^2 = 140.169$.

Figure 18.1 shows $\overline{CV}_1$ on the left, $\overline{\chi}^2$ in the middle and $\overline{CV}_2$ on the right. The height of the intersection between $\overline{CV}_1$ and $\overline{\chi}^2$, and between $\overline{\chi}^2$ and $\overline{CV}_2$, are both below 0.8. We conclude $\overline{CV}_1 < \overline{\chi}^2 < \overline{CV}_2$ and we do not reject H_0. The Maple [2] commands for Figure 18.1 are in Chapter 30. The crisp test also concludes do not reject H_0.

Example 18.3.2

Assume everything is the same as in Example 18.3.1 except that $s^2 = 2.675$, Then $\chi_0^2 = 133.75$. The graphs of all three fuzzy numbers are shown in

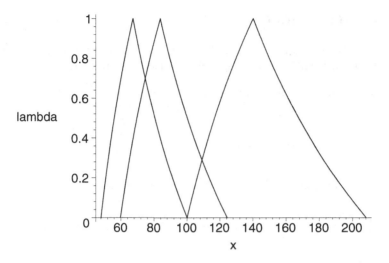

Figure 18.1: Fuzzy Test in Example 18.3.1($\overline{CV}_1$ left, $\overline{\chi}^2$ middle, $\overline{CV}_2$ right)

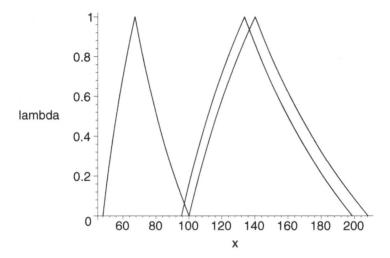

Figure 18.2: Fuzzy Test in Example 18.3.2($\overline{CV}_1$ left, $\overline{\chi}^2$ middle, $\overline{CV}_2$ right)

Figure 18.2. In Figure 18.2 $\overline{\chi}^2$ and $\overline{CV}_2$ are very close together, with $\overline{\chi}^2$ just to the left of $\overline{CV}_2$, so $\overline{CV}_2 \approx \overline{\chi}^2$. The final result is $\overline{CV}_1 < \overline{\chi}^2 \approx \overline{CV}_2$ and we have no decision on H_0. The crisp test would decide to not reject H_0.

18.4   References

1. R.V. Hogg and E.A. Tanis: Probability and Statistical Inference, Sixth Edition, Prentice Hall, Upper Saddle River, N.J., 2001.

2. Maple 9, Waterloo Maple Inc., Waterloo, Canada.

Chapter 19

Fuzzy Correlation

19.1 Introduction

We first present the crisp theory and then the fuzzy results. We have taken the crisp theory from Section 8.8 of [1].

19.2 Crisp Results

Let random variables X and Y have a bivariate normal distribution with parameters μ_x, μ_y, σ_x^2, σ_y^2 and ρ the (linear) correlation coefficient. We want to first estimate ρ and then perform the test

$$H_0 : \rho = 0, \tag{19.1}$$

verses

$$H_1 : \rho \neq 0. \tag{19.2}$$

To estimate ρ we obtain a random sample (x_i, y_i), $1 \leq i \leq n$, from the bivariate distribution. R, the sample correlation coefficient, is the point estimator of ρ and it is computed as follows

$$R = s_{xy}/s_x s_y, \tag{19.3}$$

where

$$s_{xy} = [\sum_{i=1}^{n}(x_i - \overline{x})(y_i - \overline{y})]/(n-1), \tag{19.4}$$

$$s_x = \sqrt{[\sum_{i=1}^{n}(x_i - \overline{x})^2]/(n-1)}, \tag{19.5}$$

James J. Buckley: *Fuzzy Probability and Statistics*, StudFuzz **196**, 167–170 (2006)
www.springerlink.com

$$s_y = \sqrt{[\sum_{i=1}^{n}(y_i - \overline{y})^2]/(n-1)}. \tag{19.6}$$

To obtain a confidence interval for ρ we make the transformation

$$W(R) = 0.5\ln\frac{1+R}{1-R}, \tag{19.7}$$

which has an approximate normal distribution with mean $W(\rho)$ and standard deviation $\sqrt{\frac{1}{n-3}}$. This leads to finding a $(1-\beta)100\%$ confidence interval for ρ. Let the confidence interval be written $[\rho_1(\beta), \rho_2(\beta)]$. Then

$$\rho_1(\beta) = \frac{1+R-(1-R)\exp(s)}{1+R+(1-R)\exp(s)}, \tag{19.8}$$

$$\rho_2(\beta) = \frac{1+R-(1-R)\exp(-s)}{1+R+(1-R)\exp(-s)}, \tag{19.9}$$

$$s = 2z_{\beta/2}/\sqrt{n-3}. \tag{19.10}$$

To perform the hypothesis test we determine the test statistic

$$z_0 = \sqrt{n-3}[W(R) - W(0)], \tag{19.11}$$

because under H_0 $\rho = 0$ and then $W(0) = 0$ also. So, $z_0 = \sqrt{n-3}W(R)$ has an approximate $N(0,1)$ distribution. Let γ, $0 < \gamma < 1$, be the significance level of the test. Usual values for γ are $0.10, 0.05, 0.01$. Now under the null hypothesis H_0 z_0 is $N(0,1)$ and our decision rule is: (1) reject H_0 if $z_0 \geq z_{\gamma/2}$ or $z_0 \leq -z_{\gamma/2}$; and (2) do not reject H_0 when $-z_{\gamma/2} < z_0 < z_{\gamma/2}$.

19.3 Fuzzy Theory

The first thing we want to do is construct our fuzzy estimator for ρ. We just place the $(1-\beta)100\%$ confidence intervals given in equations (19.8)-(19.10), one on top of another to get $\overline{\rho}$.

Example 19.3.1

Let the data be n=16 with computed $R = 0.35$. This is enough to get the graph of the fuzzy estimator $\overline{\rho}$ which is shown in Figure 19.1. The Maple [2] commands for this figure are in Chapter 30.

Next we go to the fuzzy test statistic. Substitute α-cuts of $\overline{\rho}$ for R in equation (19.11), assuming all intervals are positive and simply using interval arithmetic, we obtain alpha-cuts of our fuzzy statistic $\overline{Z}$

$$\overline{Z}[\alpha] = \frac{\sqrt{n-3}}{2}[ln\frac{1+\rho_1(\alpha)}{1-\rho_1(\alpha)}, ln\frac{1+\rho_2(\alpha)}{1-\rho_2(\alpha)}]. \tag{19.12}$$

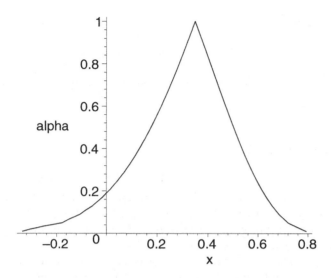

Figure 19.1: Fuzzy Estimator $\overline{\rho}$ in Example 19.3.1

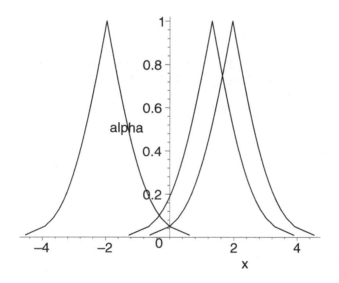

Figure 19.2: Fuzzy Test $\overline{Z}$ verses the $\overline{CV}_i$ in Example 19.3.2($\overline{CV}_1$ left, $\overline{Z}$ middle, $\overline{CV}_2$ right)

Now substitute the expressions for the $\rho_i(\alpha)$ into the above equation and simplify giving the surprising result

$$\overline{Z}[\alpha] = [z_0 - z_{\alpha/2}, z_0 + z_{\alpha/2}]. \tag{19.13}$$

Using this result we easily determine that

$$\overline{CV}_1[\alpha] = [-z_{\gamma/2} - z_{\alpha/2}, -z_{\gamma/2} + z_{\alpha/2}], \tag{19.14}$$

$$\overline{CV}_2[\alpha] = [z_{\gamma/2} - z_{\alpha/2}, z_{\gamma/2} + z_{\alpha/2}]. \tag{19.15}$$

Example 19.3.2

Use the same data as in Example 19.3.1 and let $\gamma = 0.05$ so that $\pm z_{\gamma/2} = \pm 1.96$ and we compute $z_0 = 1.3176$. The graphs of $\overline{Z}$ and the $\overline{CV}_i$ are shown in Figure 19.2. This figure shows that $\overline{CV}_1 < \overline{Z} < \overline{CV}_2$ since the heights of the intersections are both less than 0.8. Hence we do not reject H_0. The crisp test will produce the same result.

If we reject H_0 we believe that we have significant linear correlation between X and Y and then proceed on to the next three chapters to investigate this linear relationship.

19.4   References

1. R.V. Hogg and E.A. Tanis: Probability and Statistical Inference, Sixth Edition, Prentice Hall, Upper Saddle River, N.J., 2001.

2. Maple 9, Waterloo Maple Inc., Waterloo, Canada

Chapter 20

Estimation in Simple Linear Regression

20.1   Introduction

Let us first review the basic theory on crisp simple linear regression. Our development, throughout this chapter and the next two chapters, follows Sections 7.8 and 7.9 in [1]. We have some data (x_i, y_i), $1 \le i \le n$, on two variables x and Y. Notice that we start with crisp data and not fuzzy data. Most papers on fuzzy regression assume fuzzy data. The values of x are known in advance and Y is a random variable. We assume that there is no uncertainty in the x data. We can not predict a future value of Y with certainty so we decide to focus on the mean of Y, $E(Y)$. We assume that $E(Y)$ is a linear function of x, say $E(Y) = a + b(x - \overline{x})$. Here $\overline{x}$ is the mean of the x-values and not a fuzzy set. Our model is

$$Y_i = a + b(x_i - \overline{x}) + \epsilon_i, \tag{20.1}$$

where ϵ_i are independent and $N(0, \sigma^2)$ with σ^2 unknown. The basic regression equation for the mean of Y is $y = a + b(x - \overline{x})$ and now we wish to estimate the values of a and b. Notice our basic regression line is not $y = a + bx$, and the expression for the estimator of a will differ between the two models.

We will need the $(1 - \beta)100\%$ confidence interval for a and b. First we require the crisp point estimators of a, b and σ^2. The crisp estimator of a is $\widehat{a} = \overline{y}$ the mean of the y_i values. Next $\widehat{b}$ is $B1/B2$ where

$$B1 = \sum_{i=1}^{n} y_i(x_i - \overline{x}), \tag{20.2}$$

$$B2 = \sum_{i=1}^{n} (x_i - \overline{x})^2. \tag{20.3}$$

James J. Buckley: *Fuzzy Probability and Statistics*, StudFuzz **196**, 1 7 1 – 1 7 5 (2006)
www.springerlink.com   © Springer-Verlag Berlin Heidelberg 2006

Finally

$$\widehat{\sigma}^2 = (1/n) \sum_{i=1}^{n} [y_i - \widehat{a} - \widehat{b}(x_i - \overline{x})]^2. \tag{20.4}$$

Using these expressions we may construct confidence intervals for a and b.

20.2   Fuzzy Estimators

A $(1 - \beta)100\%$ confidence interval for a is

$$[\widehat{a} - t_{\beta/2}\sqrt{\widehat{\sigma}^2/(n-2)}, \widehat{a} + t_{\beta/2}\sqrt{\widehat{\sigma}^2/(n-2)}], \tag{20.5}$$

where $t_{\beta/2}$ is the value for a t-distribution, $n - 2$ degrees of freedom, so that the probability of exceeding it is $\beta/2$. From this expression we can build the triangular shaped fuzzy number estimator $\overline{a}$ for a by placing these confidence intervals one on top of another.

A $(1 - \beta)100\%$ confidence interval for b is

$$[\widehat{b} - t_{\beta/2}\sqrt{C1/C2}, \widehat{b} + t_{\beta/2}\sqrt{C1/C2}], \tag{20.6}$$

where

$$C1 = n\widehat{\sigma}^2, \tag{20.7}$$

and

$$C2 = (n-2) \sum_{i=1}^{n} (x_i - \overline{x})^2. \tag{20.8}$$

These confidence intervals for b will produce the fuzzy number estimator $\overline{b}$ for b.

We will also need the fuzzy estimator for σ^2 in Chapter 22. A $(1-\beta)100\%$ confidence interval for σ^2 is

$$[\frac{n\widehat{\sigma}^2}{\chi^2_{R,\beta/2}}, \frac{n\widehat{\sigma}^2}{\chi^2_{L,\beta/2}}], \tag{20.9}$$

where $\chi^2_{R,\beta/2}$ $(\chi^2_{L,\beta/2})$ is the point on the right (left) side of the χ^2 density where the probability of exceeding (being less than) it is $\beta/2$. This χ^2 distribution has $n - 2$ degrees of freedom. Put these confidence intervals together and we obtain $\overline{\sigma}^2$ our fuzzy number estimator of σ^2. However, as discussed in Chapter 9, this fuzzy estimator is biased. It is biased because when we evaluate at $\beta = 1$ we should obtain the point estimator $\widehat{\sigma}^2$ but we do not get this value. So to get an unbiased fuzzy estimator we will define new functions $L(\lambda)$ and $R(\lambda)$ similar to those (equations (9.8) and (9.9)) in Chapter 9. We will employ these definitions of $L(\lambda)$ and $R(\lambda)$ in this chapter and in Chapter 22.

$$L(\lambda) = [1 - \lambda]\chi^2_{R,0.005} + \lambda n, \tag{20.10}$$

x	y
70	77
74	94
72	88
68	80
58	71
54	76
82	88
64	80
80	90
61	69

Table 20.1: Crisp Data for Example 20.2.1

$$R(\lambda) = [1 - \lambda]\chi^2_{L,0.005} + \lambda n, \tag{20.11}$$

where the degrees of freedom are $n - 2$. Then a unbiased $(1 - \beta)100\%$ confidence interval for σ^2 is

$$[\frac{n\widehat{\sigma}^2}{L(\lambda)}, \frac{n\widehat{\sigma}^2}{R(\lambda)}], \tag{20.12}$$

for $0 \le \lambda \le 1$. If we evaluate this confidence interval at $\lambda = 1$ we obtain $[\widehat{\sigma}^2, \widehat{\sigma}^2] = \widehat{\sigma}^2$. Now β (α) will be a function of λ as shown in equations (9.12) and (9.13) in Chapter 9.

Example 20.2.1

The data set we will use is shown in Table 20.1 which is the data used in Example 7.8-1 in [1]. We will also use this data in the next two chapters. From this data set we compute $\widehat{a} = 81.3$, $\widehat{b} = 0.742$ and $\widehat{\sigma}^2 = 21.7709$.

Then the $(1 - \beta)100\%$ confidence interval for a is

$$[81.3 - 1.6496t_{\beta/2}, 81.3 + 1.6496t_{\beta/2}], \tag{20.13}$$

and the same confidence interval for b is

$$[0.742 - 0.1897t_{\beta/2}, 0.742 + 0.1897t_{\beta/2}], \tag{20.14}$$

and the same confidence interval for σ^2 is

$$[\frac{217.709}{L(\lambda)}, \frac{217.709}{R(\lambda)}], \tag{20.15}$$

where

$$L(\lambda) = [1 - \lambda](21.955) + 10\lambda, \tag{20.16}$$

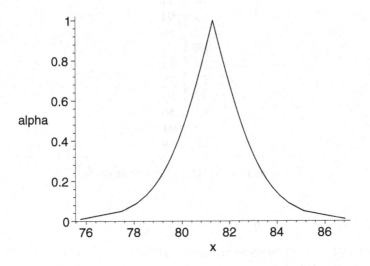

Figure 20.1: Fuzzy Estimator for a in Example 20.2.1

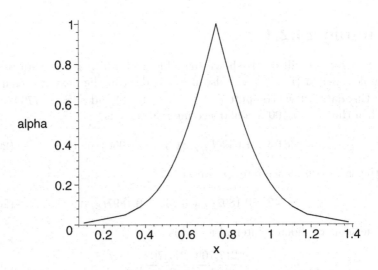

Figure 20.2: Fuzzy Estimator for b in Example 20.2.1

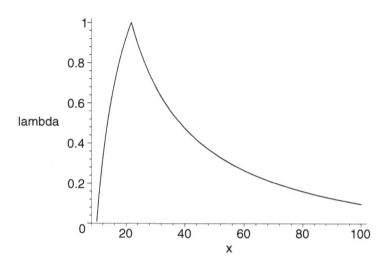

Figure 20.3: Fuzzy Estimator for σ^2 in Example 20.2.1

$$R(\lambda) = [1 - \lambda]1.344 + 10\lambda. \tag{20.17}$$

for $0 \leq \lambda \leq 1$. All degrees of freedom are 8. We graphed equations (20.13) and (20.14) as functions of $\beta = \alpha$ for $0.01 \leq \alpha \leq 1$, using Maple [2] and the results are in Figures 20.1 and 20.2. The graph of the fuzzy estimator for the variance, equation (20.15), was done for $\lambda \in [0, 1]$. Maple commands for Figure 20.3 are in Chapter 30. The fuzzy estimator for a (b, σ^2) we shall write as $\overline{a}$ $(\overline{b}, \overline{\sigma}^2)$.

Once we have these fuzzy estimators of a and b we may go on to fuzzy prediction in the next chapter.

20.3 References

1. R.V. Hogg and E.A. Tanis: Probability and Statistical Inference, Sixth Edition, Prentice Hall, Upper Saddle River, N.J., 2001.

2. Maple 9, Waterloo Maple Inc., Waterloo, Canada.

Chapter 21

Fuzzy Prediction in Linear Regression

21.1 Prediction

From the previous chapter we have our fuzzy regression equation

$$\overline{y}(x) = \overline{a} + \overline{b}(x - \overline{x}), \tag{21.1}$$

for $\overline{y}(x)$, with $\overline{a}$ and $\overline{b}$ fuzzy numbers and x and $\overline{x}$ real numbers. $\overline{y}(x)$ is our fuzzy number estimator for the mean of Y ($E(Y)$) given x, and we show this dependence on x with the notation $\overline{y}(x)$. Now $\overline{x}$ is a fixed real number but we may choose new values for x to predict new fuzzy values for $E(Y)$.

Let $\overline{a}[\alpha] = [a_1(\alpha), a_2(\alpha)]$, $\overline{b}[\alpha] = [b_1(\alpha), b_2(\alpha)]$ and $\overline{y}(x)[\alpha] = [y(x)_1(\alpha), y(x)_2(\alpha)]$. All fuzzy calculations will be done using α-cuts and interval arithmetic. The main thing to remember now from interval arithmetic (Chapter 2, Section 2.3.2) is that $c[a, b]$ equals $[ca, cb]$ if $c > 0$ but it is $[cb, ca]$ when $c < 0$. Then

$$y(x)_1(\alpha) = a_1(\alpha) + (x - \overline{x})b_1(\alpha), \tag{21.2}$$

when $(x - \overline{x}) > 0$ and

$$y(x)_1(\alpha) = a_1(\alpha) + (x - \overline{x})b_2(\alpha), \tag{21.3}$$

if $(x - \overline{x}) < 0$. Similarly

$$y(x)_2(\alpha) = a_2(\alpha) + (x - \overline{x})b_2(\alpha), \tag{21.4}$$

when $(x - \overline{x}) > 0$ and

$$y(x)_2(\alpha) = a_2(\alpha) + (x - \overline{x})b_1(\alpha), \tag{21.5}$$

if $(x - \overline{x}) < 0$. The alpha-cuts of $\overline{a}$ and $\overline{b}$ were determined in the previous chapter. There the α-cut is the $(1 - \alpha)100\%$ confidence interval.

James J. Buckley: *Fuzzy Probability and Statistics*, StudFuzz **196**, 177–179 (2006)
www.springerlink.com   © Springer-Verlag Berlin Heidelberg 2006

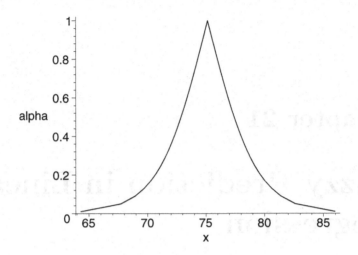

Figure 21.1: Fuzzy Estimator of $E(Y)$ given $x = 60$ in Example 21.1

Example 21.1

We use the same data as in Example 20.2.1 in Chapter 20. Here we will find $\overline{y}(60)$ and $\overline{y}(70)$. Notice that we are using $x = 70$ which is already in the data set in Table 20.1. First consider $x = 60$. Then $(x - \overline{x}) = -8.3 < 0$ because $\overline{x} = 68.3$. We use equations (21.3) and (21.5). Using Maple [2] the graph of $\overline{y}(60)$ is in Figure 21.1. If $x = 70$ then $(x - \overline{x}) = 1.7 > 0$ and use equations (21.2) and (21.4). The graph of $\overline{y}(70)$ is shown in Figure 21.2. The Maple commands for the $x = 60$ case are in Chapter 30.

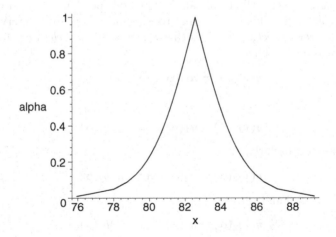

Figure 21.2: Fuzzy Estimator of $E(Y)$ given $x = 70$ in Example 21.1

Confidence Interval	$x = 60$	$x = 70$
$\overline{y}(x)[0]$	$[64.32, 85.96]$	$[75.94, 89.18]$
99% CI for $E(Y)$	$[67.49, 82.79]$	$[76.92, 88.20]$
99% CI for y	$[56.04, 94.24]$	$[64.17, 100.95]$

Table 21.1: Comparing the 99% Confidence Intervals in Example 21.1

Now let us compare these results to those obtained from probability theory. First $\overline{y}(x)[0]$ is like a 99% confidence interval for $y(x)$ because it uses $\overline{a}[0]$ ($\overline{b}[0]$) which is a 99% confidence interval for a (b). So we will compare these $\alpha = 0$ cuts to: (1) the 99% confidence interval for the mean of Y ($E(Y)$) given $x = 60(70)$; and (2) the 99% confidence interval for a value of y given $x = 60$ (70). Expressions for both of these crisp confidence intervals may be found in the statistics book [1], so we will not reproduce them here. The results are in Table 21.1 where "CI" denotes "confidence interval". Notice that in Table 21.1: (1) the 99% confidence interval for $E(Y)$ is a subset of $y(x)[0]$ for both $x = 60$ and $x = 70$; and (2) the 99% confidence interval for a value of y contains the interval $y(x)[0]$ for $x = 60, 70$. We know from crisp statistics that the confidence interval for $E(Y)$ will always be a subset of the confidence interval for a value of y. However, we do not always expect, for all other data sets, $y(x)[0]$ to be between these other two intervals.

21.2　References

1. R.V. Hogg and E.A. Tanis: Probability and Statistical Inference, Sixth Edition, Prentice Hall, Upper Saddle River, N.J., 2001.

2. Maple 9, Waterloo Maple Inc., Waterloo, Canada.

Chapter 22

Hypothesis Testing in Regression

22.1   Introduction

We look at two fuzzy hypothesis tests in this chapter: (1) in the next section $H_0 : a = a_0$ verses $H_1 : a > a_0$ a one-sided test; and (2) in the following section $H_0 : b = 0$ verses $H_1 : b \neq 0$. In both cases we first review the crisp (non-fuzzy) test before the fuzzy test. The non-fuzzy hypothesis tests are based on Sections 7.8 and 7.9 of [1].

22.2   Tests on a

Let us first review the crisp situation. We wish to do the following hypothesis test

$$H_0 : a = a_0, \tag{22.1}$$

verses

$$H_1 : a > a_0, \tag{22.2}$$

which is a one-sided test. Then we determine the statistic

$$t_0 = \frac{\widehat{a} - a_0}{\sqrt{\widehat{\sigma}^2/(n-2)}}, \tag{22.3}$$

which, under H_0, has a t distribution with $(n-2)$ degrees of freedom. Let γ, $0 < \gamma < 1$, be the significance level of the test. Usual values for γ are $0.10, 0.05, 0.01$. Our decision rule is: (1) reject H_0 if $t_0 \geq t_\gamma$; and (2) do not reject H_0 when $t_0 < t_\gamma$. In the above decision rule t_γ is the t-value so that the probability of a random variable, having the t probability density, exceeding t is γ. The critical region is $[t_\gamma, \infty)$ with critical value t_γ.

James J. Buckley: *Fuzzy Probability and Statistics*, StudFuzz **196**, 181–185 (2006)
www.springerlink.com   © Springer-Verlag Berlin Heidelberg 2006

Now proceed to the fuzzy situation where our fuzzy estimator of a is the triangular shaped fuzzy number $\overline{a}$ developed in Chapter 20. We will also need the fuzzy estimator for σ^2 also given in Chapter 20. Then our fuzzy statistic is

$$\overline{T} = \frac{\overline{a} - a_0}{\sqrt{\overline{\sigma}^2/(n-2)}}. \tag{22.4}$$

All fuzzy calculations will be performed via α-cuts and interval arithmetic. We find, after substituting the intervals for an alpha-cut of $\overline{a}$ and $\overline{\sigma}^2$ into the expression for $\overline{T}$, using interval arithmetic, and simplification, that

$$\overline{T}[\alpha] = [\Pi_1(t_0 - t_{\alpha/2}), \Pi_2(t_0 + t_{\alpha/2})], \tag{22.5}$$

where

$$\Pi_1 = \sqrt{R(\lambda)/n}, \tag{22.6}$$

and

$$\Pi_2 = \sqrt{L(\lambda)/n}. \tag{22.7}$$

$L(\lambda)$ and $R(\lambda)$ were defined in equations (20.10) and (20.11), respectively, in Chapter 20.

We have assumed that all intervals are positive in the derivation of equation (22.5). The interval for an alpha-cut of $\overline{a}$ may be positive or negative, but the interval for an alpha-cut of $\overline{\sigma}^2$ is always positive. When the left end point (or both end points) of the interval for an alpha-cut of $\overline{a}$ is negative we have to make some changes in equation (22.5). See section 16.3.1 in Chapter 16 for the details.

Now that we know the alpha-cuts of the fuzzy statistic we can find α-cuts of the fuzzy critical value $\overline{CV}_2$ for this one-sided test. As in previous chapters we get

$$\overline{CV}_2[\alpha] = [\Pi_1(t_\gamma - t_{\alpha/2}), \Pi_2(t_\gamma + t_{\alpha/2})]. \tag{22.8}$$

In this equation γ is fixed and alpha varies in the interval $[0.01, 1]$.

We now have a fuzzy set $\overline{T}$ for our test statistic and a fuzzy set $\overline{CV}_2$ for the critical value. Our final decision will depend on the relationship between $\overline{T}$ and $\overline{CV}_2$. Our test becomes : (1) reject H_0 if $\overline{T} > \overline{CV}_2$; (2) do not reject if $\overline{T} < \overline{CV}_2$; and (3) there is no decision on H_0 if $\overline{T} \approx \overline{CV}_2$.

Example 22.2.1

We will still use the data in Table 20.1 and we have computed $\widehat{a} = 81.3$, $\widehat{b} = 0.742$ and $\widehat{\sigma}^2 = 21.7709$ with $n = 10$. Let $\gamma = 0.05$, $a_0 = 80$ and determine $t_0 = 0.7880$ and $t_{0.05} = 1.860$ with 8 degrees of freedom. We compute

$$L(\lambda) = 21.955 - 11.955\lambda, \tag{22.9}$$

$$R(\lambda) = 1.344 + 8.656\lambda, \tag{22.10}$$

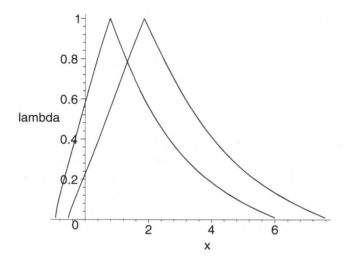

Figure 22.1: Fuzzy Test $\overline{T}$ verses $\overline{CV}_2$ in Example 22.2.1($\overline{T}$ left, $\overline{CV}_2$ right)

$$\Pi_1 = \sqrt{0.1344 + 0.8656\lambda}, \tag{22.11}$$

$$\Pi_2 = \sqrt{2.1955 - 1.1955\lambda}. \tag{22.12}$$

From these results we may get the graphs of $\overline{T}$ and $\overline{CV}_2$ and they are shown in Figure 22.1. The Maple [2] commands for this figure are in Chapter 30.

From Figure 22.1 we see that $\overline{T} < \overline{CV}_2$ since the height of the intersection is less than 0.8. We therefore conclude: do not reject H_0. Of course, the crisp test would have the same result.

The graph of $\overline{T}$ in Figure 22.1 is not entirely correct on its left side to the left of the vertical axis. This is because in computing α-cuts of $\overline{T}$ the left end point of the interval for the numerator (see Section 16.3.1) goes negative for $0 \le \alpha < \alpha^*$ and we had assumed it was always positive. Also, this effects the left side of $\overline{CV}_2$ for $0 \le \alpha < \alpha^*$. However, these changes do not effect the final decision because it depends on comparing the right side of $\overline{T}$ to the left side of $\overline{CV}_2$.

22.3   Tests on b

Let us first discuss the crisp hypothesis test. We wish to do the following hypothesis test

$$H_0 : b = 0, \tag{22.13}$$

verses

$$H_1 : b \neq 0, \tag{22.14}$$

which is a two-sided test. Next determine the statistic

$$t_0 = \frac{\widehat{b} - 0}{\sqrt{d\widehat{\sigma}^2/(n-2)}}, \tag{22.15}$$

where

$$d = \frac{n}{\sum_{i=1}^{n}(x_i - \overline{x})^2}, \tag{22.16}$$

which, under H_0, t_0 has a t distribution with $(n-2)$ degrees of freedom. Let γ, $0 < \gamma < 1$, be the significance level of the test. Our decision rule is: (1) reject H_0 if $t_0 \geq t_{\gamma/2}$ or if $t_0 \leq -t_{\gamma/2}$; and (2) otherwise do not reject H_0.

Now proceed to the fuzzy situation where our estimate of b is the triangular shaped fuzzy number $\overline{b}$ and our fuzzy estimator $\overline{\sigma}^2$ of σ^2 is also a fuzzy number. These fuzzy estimators were deduced in Chapter 20. Then our fuzzy statistic is

$$\overline{T} = \frac{\overline{b} - 0}{\sqrt{d\overline{\sigma}^2/(n-2)}}. \tag{22.17}$$

All fuzzy calculations will be performed via α-cuts and interval arithmetic. We find, after substituting the intervals for an alpha-cuts of $\overline{b}$ and $\overline{\sigma}^2$ into the expression for $\overline{T}$, using interval arithmetic, assuming all intervals are positive, that

$$\overline{T}[\alpha] = [\Pi_1(t_0 - t_{\alpha/2}), \Pi_2(t_0 + t_{\alpha/2})], \tag{22.18}$$

where

$$\Pi_1 = \sqrt{R(\lambda)/n}, \tag{22.19}$$

and

$$\Pi_2 = \sqrt{L(\lambda)/n}. \tag{22.20}$$

$L(\lambda)$ and $R(\lambda)$ were defined in equations (20.10) and (20.11) in Chapter 20.

Now that we know the alpha-cuts of the fuzzy statistic we can find α-cuts of the fuzzy critical values $\overline{CV}_i$, $i = 1, 2$. As in previous chapters we obtain

$$\overline{CV}_1[\alpha] = [\Pi_2(-t_{\gamma/2} - t_{\alpha/2}), \Pi_1(-t_{\gamma/2} + t_{\alpha/2})], \tag{22.21}$$

and because $\overline{CV}_2 = -\overline{CV}_1$

$$\overline{CV}_2[\alpha] = [\Pi_1(t_{\gamma/2} - t_{\alpha/2}), \Pi_2(t_{\gamma/2} + t_{\alpha/2})]. \tag{22.22}$$

In these equations γ is fixed and alpha varies in the interval $[0.01, 1]$.

Given the fuzzy numbers $\overline{T}$ and the $\overline{CV}_i$, $i = 1, 2$, we may compare $\overline{T}$ to $\overline{CV}_1$ and then to $\overline{CV}_2$ to determine our final conclusion on H_0.

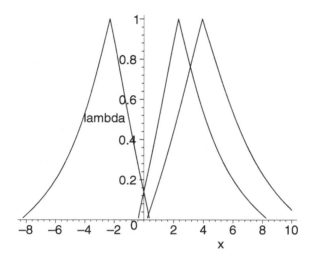

Figure 22.2: Fuzzy Test $\overline{T}$ verses the $\overline{CV}_i$ in Example 22.3.1 ($\overline{CV}_1$ left, $\overline{CV}_2$ center, $\overline{T}$ right)

Example 22.3.1

We will still use the data in Table 20.1 and we have already computed $\widehat{b} = 0.742$, and $\widehat{\sigma}^2 = 21.7709$ with $n = 10$. Let $\gamma = 0.05$, and compute $t_0 = 3.9111$ and $t_{0.025} = 2.306$ with 8 degrees of freedom.

The values of $L(\lambda)$, $R(\lambda)$, Π_1 and Π_2 are all the same as in Example 22.2.1. All that has changed is the value of t_0 and that now we use both $\overline{CV}_1$ and $\overline{CV}_2$ for a two-sided test.

The graphs of $\overline{T}$ and the $\overline{CV}_i$ are shown in Figure 22.2. It is evident that $\overline{CV}_2 < \overline{T}$, because the height of the intersection is less than 0.8. Hence we reject H_0, the same as in the crisp case.

22.4 References

1. R.V. Hogg and E.A. Tanis: Probability and Statistical Inference, Sixth Edition, Prentice Hall, Upper Saddle River, N.J., 2001.

2. Maple 9, Waterloo Maple Inc., Waterloo, Canada.

Chapter 23

Estimation in Multiple Regression

23.1 Introduction

We first review the basic theory on crisp multiple linear regression. For simplicity let us work with only two independent variables x_1 and x_2. Our development, throughout this chapter, and the next two chapters, follows [1]. We have some data (x_{1i}, x_{2i}, y_i), $1 \leq i \leq n$, on three variables x_1, x_2 and Y. Notice that we start with crisp data and not fuzzy data. The values of x_1 and x_2 are known in advance and Y is a random variable. We assume that there is no uncertainty in the x_1 and x_2 data. We can not predict a future value of Y with certainty so we decide to focus on the mean of Y, $E(Y)$. We assume that $E(Y)$ is a linear function of x_1 and x_2, say $E(Y) = a + bx_1 + cx_2$. Our model is

$$Y_i = a + bx_{1i} + cx_{2i} + \epsilon_i, \tag{23.1}$$

$1 \leq i \leq n$. The basic regression equation for the mean of Y is $y = a + bx_1 + cx_2$ and now we wish to estimate the values of a, b and c.

We will need the $(1 - \beta)100\%$ confidence interval for a, b and c. First, we require the crisp point estimators of a, b, c and σ^2. It is best now to turn to matrix notation in order to describe the estimators and their confidence intervals.

Let vector $\theta = [a, b, c]$, vector $\epsilon = [\epsilon_1, ..., \epsilon_n]$ and vector $y = [y_1, ..., y_n]$. If v is a $1 \times n$ vector we will use the notation v^t for the transpose of v. Then v^t is a $n \times 1$ vector. Define the $n \times 3$ matrix X as

$$X = \begin{pmatrix} 1 & x_{11} & x_{21} \\ 1 & x_{12} & x_{22} \\ .. & .. & .. \\ .. & .. & .. \\ 1 & x_{1n} & x_{2n} \end{pmatrix}. \tag{23.2}$$

James J. Buckley: *Fuzzy Probability and Statistics*, StudFuzz **196**, 187−192 (2006)
www.springerlink.com   © Springer-Verlag Berlin Heidelberg 2006

Next define $\widehat{\theta} = [\widehat{a}, \widehat{b}, \widehat{c}]$ the vector of point estimates of a, b, c. Then

$$\widehat{\theta}^t = (X^t X)^{-1} X^t y^t, \tag{23.3}$$

which gives $\widehat{a}, \widehat{b}$ and $\widehat{c}$.

The distribution of ϵ is also needed. We know that the expected value of $\epsilon^t \epsilon$ is $\sigma^2 I$ for 3×3 identity matrix I and unknown variance σ^2. In the next two chapters a point estimate for σ^2, and confidence intervals, are required. A point estimator $\widehat{\sigma}^2$ for σ^2 is

$$\widehat{\sigma}^2 = \sum_{i=1}^{n} e_i^2 / (n - 3), \tag{23.4}$$

where

$$\sum_{i=1}^{n} e_i^2 = \sum_{i=1}^{n} (y_i - \widehat{y}_i)^2, \tag{23.5}$$

$$\widehat{y}_i = \widehat{a} + \widehat{b} x_{1i} + \widehat{c} x_{2i}. \tag{23.6}$$

23.2   Fuzzy Estimators

Now we may find the confidence intervals for a, b, c and σ^2. Let $(X^t X)^{-1} = A = [a_{ij}]$. A $(1 - \beta)100\%$ confidence interval for a is

$$[\widehat{a} - t_{\beta/2} \widehat{\sigma} \sqrt{a_{11}}, \widehat{a} + t_{\beta/2} \widehat{\sigma} \sqrt{a_{11}}], \tag{23.7}$$

for a_{11} the first element along the main diagonal of matrix A. Then a $(1 - \beta)100\%$ confidence interval for b is

$$[\widehat{b} - t_{\beta/2} \widehat{\sigma} \sqrt{a_{22}}, \widehat{a} + t_{\beta/2} \widehat{\sigma} \sqrt{a_{22}}], \tag{23.8}$$

and for c

$$[\widehat{c} - t_{\beta/2} \widehat{\sigma} \sqrt{a_{33}}, \widehat{a} + t_{\beta/2} \widehat{\sigma} \sqrt{a_{33}}]. \tag{23.9}$$

In the t distribution, to find the critical value $t_{\beta/2}$, we use $n - 3$ degrees of freedom. Now put these confidence intervals together, one on top of another, to get the fuzzy estimators $\overline{a}, \overline{b}, \overline{c}$ of a, b, c, respectively.

The next item we need is a confidence interval for σ^2. It is known that

$$(n - 3)\widehat{\sigma}^2 / \sigma^2, \tag{23.10}$$

has a chi-square distribution with $n - 3$ degrees of freedom. Then

$$P(\chi_{L,\beta/2}^2 < \frac{(n - 3)\widehat{\sigma}^2}{\sigma^2} < \chi_{R,\beta/2}^2) = 1 - \beta, \tag{23.11}$$

and if we solve this equation for σ^2, it leads to the $(1 - \beta)100\%$ confidence interval for σ^2, which is

$$[\frac{(n - 3)\widehat{\sigma}^2}{\chi_{R,\beta/2}^2}, \frac{(n - 3)\widehat{\sigma}^2}{\chi_{L,\beta/2}^2}]. \tag{23.12}$$

where $\chi^2_{R,\beta/2}$ $(\chi^2_{L,\beta/2})$ is the point on the right (left) side of the χ^2 density where the probability of exceeding (being less than) it is $\beta/2$. This χ^2 distribution has $n-3$ degrees of freedom. Put these confidence intervals together and we obtain $\overline{\sigma}^2$ our fuzzy number estimator of σ^2. However, as discussed in Chapter 9 and 20, this fuzzy estimator is biased. It is biased because when we evaluate at $\beta = 1$ we should obtain the point estimator $\widehat{\sigma}^2$ but we do not get this value. So to get an unbiased fuzzy estimator we will define new functions $L(\lambda)$ and $R(\lambda)$, similar to those in Chapter 9. We will employ these definitions of $L(\lambda)$ and $R(\lambda)$ in this chapter and in Chapter 25.

$$L(\lambda) = [1 - \lambda]\chi^2_{R,0.005} + \lambda(n-3), \tag{23.13}$$

$$R(\lambda) = [1 - \lambda]\chi^2_{L,0.005} + \lambda(n-3), \tag{23.14}$$

where the degrees of freedom are $n-3$. Then a unbiased $(1-\beta)100\%$ fuzzy estimator for σ^2 is $\overline{\sigma}^2$ whose α-cuts are

$$[\frac{(n-3)\widehat{\sigma}^2}{L(\lambda)}, \frac{(n-3)\widehat{\sigma}^2}{R(\lambda)}], \tag{23.15}$$

for $0 \leq \lambda \leq 1$. If we evaluate this confidence interval at $\lambda = 1$ we obtain $[\widehat{\sigma}^2, \widehat{\sigma}^2] = \widehat{\sigma}^2$. Now β will be a function of λ as shown in equations (9.12) and (9.13) in Chapter 9.

Example 23.2.1

The data we shall use is in Table 23.1 which is from an example in [1]. This same data will be in the examples in the next two chapters. We want to construct the graphs of the fuzzy estimators.

We first, using Maple [2], computed the point estimators and obtained $\widehat{a} = -49.3413$, $\widehat{b} = 1.3642$, $\widehat{c} = 0.1139$ and $\widehat{\sigma}^2 = 12.9236$. Next we found

Y	x_1	x_2
100	100	100
106	104	99
107	106	110
120	111	126
110	111	113
116	115	103
123	120	102
133	124	103
137	126	98

Table 23.1: Crisp Data for Example 23.2.1

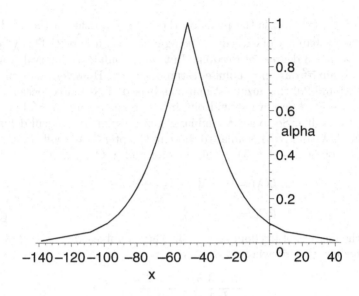

Figure 23.1: Fuzzy Estimator $\bar{a}$ for a in Example 23.2.1

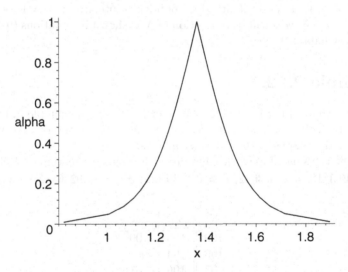

Figure 23.2: Fuzzy Estimator $\bar{b}$ for b in Example 23.2.1

the values down the main diagonal of $(X^t X)^{-1}$ and they were $a_{11} = 44.7961$, $a_{22} = 0.001586$ and $a_{33} = 0.001591$. The equations that determine the alpha-cuts of the fuzzy estimators are

$$-49.3413 \pm t_{\alpha/2}(24.0609), \qquad (23.16)$$

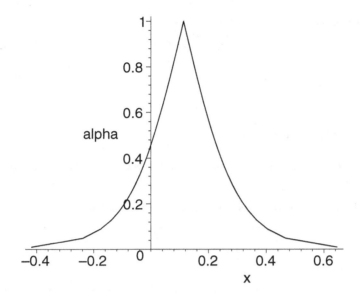

Figure 23.3: Fuzzy Estimator $\bar{c}$ for c in Example 23.2.1

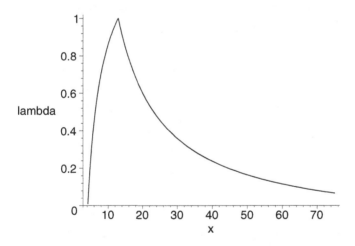

Figure 23.4: Fuzzy Estimator $\bar{\sigma}^2$ for σ^2 in Example 23.2.1

for $\bar{a}$, and for $\bar{b}$ it is

$$1.3642 \pm t_{\alpha/2}(0.1432), \qquad (23.17)$$

and

$$0.1139 \pm t_{\alpha/2}(0.1434), \qquad (23.18)$$

goes with $\bar{c}$, and

$$[\frac{77.5416}{L(\lambda)}, \frac{77.5416}{R(\lambda)}], \tag{23.19}$$

for $\bar{\sigma}^2$. These are all graphed, $0.01 \leq \alpha \leq 1$, and the results are in Figures 23.1-23.4.

23.3   References

1. J. Johnston: Econometric Methods, Second Edition, McGraw-Hill, N.Y., 1972.

2. Maple 9, Waterloo Maple Inc., Waterloo, Canada.

Chapter 24

Fuzzy Prediction in Regression

24.1 Prediction

From the previous chapter we have our fuzzy regression equation

$$\overline{y}(x_1, x_2) = \overline{a} + \overline{b}x_1 + \overline{c}x_2, \tag{24.1}$$

for $\overline{y}(x_1, x_2)$, with $\overline{a}$, $\overline{b}$ and $\overline{c}$ fuzzy numbers and x_1, x_2 real numbers. $\overline{y}(x_1, x_2)$ is our fuzzy number estimator for the mean of Y ($E(Y)$) given x_1 and x_2, and we show this dependence on x_1 and x_2 with the notation $\overline{y}(x_1, x_2)$. We may choose new values for x_1 and x_2 to predict new fuzzy values for $E(Y)$.

Let $\overline{a}[\alpha] = [a_1(\alpha), a_2(\alpha)]$, $\overline{b}[\alpha] = [b_1(\alpha), b_2(\alpha)]$, $\overline{c}[\alpha] = [c_1(\alpha), c_2(\alpha)]$ and $\overline{y}(x_1, x_2)[\alpha] = [y(x_1, x_2)_1(\alpha), y(x_1, x_2)_2(\alpha)]$. All fuzzy calculations will be done using α-cuts and interval arithmetic. Now from Example 23.2.1, and the data in Table 23.1, we assume the new values of x_1 and x_2 are positive. The only thing to remember from interval arithmetic (Chapter 2, Section 2.3.2) is that $e[a, b]$ equals $[ea, eb]$ when $e > 0$. Then

$$y(x_1, x_2)_1(\alpha) = a_1(\alpha) + x_1 b_1(\alpha) + x_2 c_1(\alpha), \tag{24.2}$$

and

$$y(x_1, x_2)_2(\alpha) = a_2(\alpha) + x_1 b_2(\alpha) + x_2 c_2(\alpha), \tag{24.3}$$

for all $\alpha \in [0, 1]$. The alpha-cuts of $\overline{a}$, $\overline{b}$ and $\overline{c}$ were determined in the previous chapter. There the α-cut is the $(1 - \alpha)100\%$ confidence interval.

Example 24.1.1

We use the same data as in Example 23.2.1 in Chapter 23. Let us assume now that the data in Table 23.1 is yearly data with the last row corresponding to

James J. Buckley: *Fuzzy Probability and Statistics*, StudFuzz **196**, 193–195 (2006)
www.springerlink.com   © Springer-Verlag Berlin Heidelberg 2006

2005. Assuming values for x_1 and x_2 for 2006 and 2007 we wish to predict $E(Y)$ for those two future years. We will find $\overline{y}(128, 96)$ and $\overline{y}(132, 92)$.

First we graphed equations (24.2) and (24.3) using $x_1 = 128$ and $x_2 = 96$. The result is in Figure 24.1. The Maple [2] commands for this figure are in Chapter 30. Next we graphed these two equations having $x_1 = 132$ and $x_2 = 92$ which is shown in Figure 24.2.

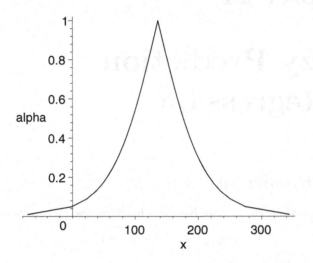

Figure 24.1: Fuzzy Estimator of $E(Y)$ Given $x_1 = 128$, $x_2 = 96$, in Example 24.1.1

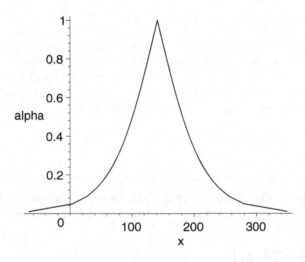

Figure 24.2: Fuzzy Estimator of $E(Y)$ Given $x_1 = 132$, $x_2 = 92$, in Example 24.1.1

Confidence Interval	$x_1 = 128, x_2 = 96$	$x_1 = 132, x_2 = 92$
$\overline{y}(x_1, x_2)[0]$	$[-71.98, 344.42]$	$[-66.97, 349.42]$
99% CI for $E(Y)$	$[126.38, 146.05]$	$[128.93, 153.49]$
99% CI for y	$[119.65, 152.77]$	$[123.09, 159.34]$

Table 24.1: Comparing the 99% Confidence Intervals in Example 24.1.1

Now let us compare these results to those obtained from probability theory. First $\overline{y}(x_1, x_2)[0]$ is like a 99% confidence interval for $y(x_1, x_2)$ because it uses $\overline{a}[0]$ ($\overline{b}[0]$, $\overline{c}[0]$) which is a 99% confidence interval for a (b, c). So we will compare these $\alpha = 0$ cuts to: (1) the 99% confidence interval for the mean of Y ($E(Y)$) given $x_1 = 128$, $x_2 = 96(x_1 = 132, x_2 = 92)$; and (2) the 99% confidence interval for a value of y given $x_1 = 128$, $x_2 = 96$ ($x_1 = 132$, $x_2 = 92$). Expressions for both of these crisp confidence intervals may be found in Section 5-5 of [1] and are reproduced below.

Let $d = (1, x_1^*, x_2^*)$ where x_i^* are new values of x_i, $i = 1, 2$. The 99% confidence interval for $E(Y)$ is

$$d\widehat{\theta}^t \pm (3.707)\widehat{\sigma}\sqrt{d(X^tX)^{-1}d^t}, \qquad (24.4)$$

where $t_{0.005} = 3.707$ for 6 degrees of freedom and the rest of the terms ($\widehat{\theta}$, $\widehat{\sigma}$,..) were defined in Chapter 23. The 99% confidence interval for the value of y is

$$d\widehat{\theta}^t \pm (3.707)\widehat{\sigma}\sqrt{1 + d(X^tX)^{-1}d^t}. \qquad (24.5)$$

The results are in Table 24.1 where "CI" denotes "confidence interval". Notice that in Table 24.1 that: (1) the 99% confidence interval for $E(Y)$ is a subset of $\overline{y}(x_1, x_2)[0]$ for both $x_1 = 128$, $x_2 = 96$ and $x_1 = 132$, $x_2 = 92$; and (2) the 99% confidence interval for a value of y is also contained in the interval $\overline{y}(x_1, x_2)[0]$ for the given new values of x_1 and x_2. We know from crisp statistics that the confidence interval for $E(Y)$ will always be a subset of the confidence interval for a value of y. However, we do not always expect, for all other data sets, $\overline{y}(x_1, x_2)[0]$ will contain the other two intervals (see Example 21.1). If fact, in this example the interval $\overline{y}(x_1, x_2)[0]$ turns out to be rather large because it combines three intervals $\overline{a}[0]$, $128(132)\overline{b}[0]$ and $96(92)\overline{c}[0]$. For example, using $x_1 = 128$ and $x_2 = 96$, then: (1) $\overline{a}[0]$ has length ≈ 178; (2) $128\overline{b}[0]$ has length ≈ 136; and (3) $96\overline{c}[0]$ has length ≈ 102. We add these lengths up we get that the length of the interval for $\overline{y}(128, 96)[0]$ is ≈ 416.

24.2 References

1. J. Johnston: Econometric Methods, Second Edition, McGraw-Hill, N.Y., 1972.

2. Maple 9, Waterloo Maple Inc., Waterloo, Canada.

Chapter 25

Hypothesis Testing in Regression

25.1 Introduction

We look at two fuzzy hypothesis tests in this chapter: (1) in the next section $H_0 : b = 0$ verses $H_1 : b > 0$ a one-sided test; and (2) in the third section $H_0 : c = 0$ verses $H_1 : c \neq 0$ a two-sided test. In both cases we first review the crisp (non-fuzzy) test before the fuzzy test. We could also runs tests on a. However, we will continue to use the data in Table 23.1 were we determined $\widehat{a} = -49.3413$, so a is definitely negative and a test like $H_0 : a = 0$ verses $H_1 : a < 0$ seems a waste of time.

25.2 Tests on b

Let us first review the crisp situation. We wish to do the following hypothesis test

$$H_0 : b = 0, \tag{25.1}$$

verses

$$H_1 : b > 0, \tag{25.2}$$

which is a one-sided test. This is a one-sided test (see also Section 15.4 of Chapter 15 and Section 22.2 of Chapter 22). Then we determine the statistic [1]

$$t_0 = \frac{\widehat{b} - 0}{\widehat{\sigma}\sqrt{a_{22}}}, \tag{25.3}$$

which, under H_0, has a t distribution with $(n - 3)$ degrees of freedom. The a_{ii}, $1 \leq i \leq 3$, are the elements on the main diagonal of $(X^t X)^{-1}$ (see Section 23.2 of Chapter 23). Let γ, $0 < \gamma < 1$, be the significance level of the test.

James J. Buckley: *Fuzzy Probability and Statistics*, StudFuzz **196**, 197–201 (2006)
www.springerlink.com

Usual values for γ are $0.10, 0.05, 0.01$. Our decision rule is: (1) reject H_0 if $t_0 \geq t_\gamma$; and (2) do not reject H_0 when $t_0 < t_\gamma$. In the above decision rule t_γ is the t-value so that the probability of a random variable, having the t probability density, exceeding t is γ. The critical region is $[t_\gamma, \infty)$ with critical value t_γ.

Now proceed to the fuzzy situation where our estimate of b is the triangular shaped fuzzy number $\bar{b}$ developed in Chapter 23. We will also need the fuzzy estimator for σ^2 also given in Chapter 23. Then our fuzzy statistic is

$$\overline{T} = \frac{\bar{b} - 0}{\overline{\sigma}\sqrt{a_{22}}}. \tag{25.4}$$

All fuzzy calculations will be performed via α-cuts and interval arithmetic. We find, after substituting the intervals for an alpha-cut of $\bar{b}$ and $\overline{\sigma}$ (square roots of equation (23.15)) into the expression for $\overline{T}$, using interval arithmetic, and simplification, that

$$\overline{T}[\alpha] = [\Pi_1(t_0 - t_{\alpha/2}), \Pi_2(t_0 + t_{\alpha/2})], \tag{25.5}$$

where

$$\Pi_1 = \sqrt{R(\lambda)/(n-3)}, \tag{25.6}$$

and

$$\Pi_2 = \sqrt{L(\lambda)/(n-3)}. \tag{25.7}$$

The $L(\lambda)$ and $R(\lambda)$ were defined in equations (23.13) and (23.14), respectively, in Chapter 23.

We have assumed that all intervals are positive in the derivation of equation (25.5). The interval for an alpha-cut of $\bar{b}$ may be positive or negative, but the interval for an alpha-cut of $\overline{\sigma}^2$ is always positive. When the left end point (or both end points) of the interval for an alpha-cut of $\bar{b}$ is negative we have to make some changes in equation (25.5). See section 16.3.1 in Chapter 16 for the details.

Now that we know the alpha-cuts of the fuzzy statistic we can find α-cuts of the fuzzy critical value $\overline{CV}_2$ for this one-sided test. As in previous chapters we get

$$\overline{CV}_2[\alpha] = [\Pi_1(t_\gamma - t_{\alpha/2}), \Pi_2(t_\gamma + t_{\alpha/2})]. \tag{25.8}$$

In this equation γ is fixed and alpha varies in the interval $[0.01, 1]$.

We now have a fuzzy set $\overline{T}$ for our test statistic and a fuzzy set $\overline{CV}_2$ for the critical value. Our final decision will depend on the relationship between $\overline{T}$ and $\overline{CV}_2$. Our test becomes : (1) reject H_0 if $\overline{T} > \overline{CV}_2$; (2) do not reject if $\overline{T} < \overline{CV}_2$; and (3) there is no decision on H_0 if $\overline{T} \approx \overline{CV}_2$.

Example 25.2.1

We will still use the data in Table 23.1 and we have computed $\hat{a} = -49.3413$, $\hat{b} = 1.3642$. $\hat{c} = 0.1139$ and $\hat{\sigma}^2 = 12.9236$ with $n = 9$ and $a_{22} = 0.001586$.

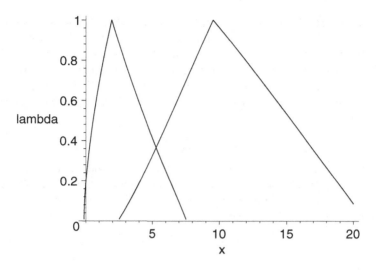

Figure 25.1: Fuzzy Test $\overline{T}$ verses $\overline{CV}_2$ in Example 25.2.1($\overline{CV}_2$ left, $\overline{T}$ right)

Let $\gamma = 0.05$, and determine $t_0 = 9.5287$ and $t_{0.05} = 1.943$ with 6 degrees of freedom. We compute

$$L(\lambda) = 18.548 - 12.548\lambda, \tag{25.9}$$

$$R(\lambda) = 0.676 + 5.324\lambda, \tag{25.10}$$

$$\Pi_1 = \sqrt{0.1127 + 0.8873\lambda}, \tag{25.11}$$

$$\Pi_2 = \sqrt{3.0913 - 2.0913\lambda}. \tag{25.12}$$

From these results we may get the graphs of $\overline{T}$ and $\overline{CV}_2$, using Maple [2], and they are shown in Figure 25.1.

From Figure 25.1 we see that $\overline{T} > \overline{CV}_2$. We therefore conclude: reject H_0. Of course, the crisp test would have the same result.

25.3 Tests on c

Let us first discuss the crisp hypothesis test. We wish to do the following hypothesis test

$$H_0 : c = 0, \tag{25.13}$$

verses

$$H_1 : c \neq 0, \tag{25.14}$$

which is a two-sided test. Next determine the statistic [1]

$$t_0 = \frac{\widehat{c} - 0}{\widehat{\sigma}\sqrt{a_{33}}}, \tag{25.15}$$

where, under H_0, t_0 has a t distribution with $(n-3)$ degrees of freedom. Let γ, $0 < \gamma < 1$, be the significance level of the test. Our decision rule is: (1) reject H_0 if $t_0 \geq t_{\gamma/2}$ or if $t_0 \leq -t_{\gamma/2}$; and (2) otherwise do not reject H_0.

Now proceed to the fuzzy situation where our estimate of c is the triangular shaped fuzzy number $\overline{c}$ and our fuzzy estimator $\overline{\sigma}^2$ of σ^2 is also a fuzzy number. These fuzzy estimators were deduced in Chapter 23. Then our fuzzy statistic is

$$\overline{T} = \frac{\overline{c} - 0}{\overline{\sigma}\sqrt{a_{33}}}. \tag{25.16}$$

All fuzzy calculations will be performed via α-cuts and interval arithmetic. We find, after substituting the intervals for an alpha-cuts of $\overline{c}$ and $\overline{\sigma}^2$ (square root of equation (23.15)) into the expression for $\overline{T}$, using interval arithmetic, assuming all intervals are positive, that

$$\overline{T}[\alpha] = [\Pi_1(t_0 - t_{\alpha/2}), \Pi_2(t_0 + t_{\alpha/2})], \tag{25.17}$$

where the Π_i were defined in the previous section.

Now that we know the alpha-cuts of the fuzzy statistic we can find α-cuts of the fuzzy critical values $\overline{CV}_i$, $i = 1, 2$. As in previous chapters we obtain

$$\overline{CV}_2[\alpha] = [\Pi_1(t_{\gamma/2} - t_{\alpha/2}), \Pi_2(t_{\gamma/2} + t_{\alpha/2})], \tag{25.18}$$

and $\overline{CV}_1 = -\overline{CV}_2$.

Given the fuzzy numbers $\overline{T}$ and the $\overline{CV}_i$, $i = 1, 2$, we may compare $\overline{T}$ to $\overline{CV}_1$ and then to $\overline{CV}_2$ to determine our final conclusion on H_0.

Example 25.3.1

We will still use the data in Table 23.1 and we have already computed $\widehat{c} = 0.1139$, $\widehat{\sigma}^2 = 12.9236$ with $n = 9$ and $a_{33} = 0.001591$. Let $\gamma = 0.05$, and compute $t_0 = 0.7943$ and $t_{0.025} = 2.447$ with 6 degrees of freedom.

The values of $L(\lambda)$, $R(\lambda)$, Π_1 and Π_2 are all the same as in Example 25.2.1. All that has changed is the value of t_0 and that now we use both $\overline{CV}_1$ and $\overline{CV}_2$ for a two-sided test.

The graphs of $\overline{T}$ and the $\overline{CV}_i$ are shown in Figure 25.2. It is evident that $\overline{CV}_2 > \overline{T}$, because the height of the intersection is less than 0.8. The point of intersection is close to 0.8, but slightly less than 0.8. Now compare $\overline{T}$ to $\overline{CV}_1$. The graph of $\overline{T}$ is not correct to the left of the vertical axis because in computing the alpha-cuts of $\overline{T}$ we had assumed that all intervals were always positive, which is not true. However, there is no need to correct this because even if we did the height of the intersection of $\overline{T}$ and $\overline{CV}_1$ would be less than 0.8. You see that the left side of $\overline{T}$ crosses the vertical axis below 0.8 so it must, even with corrections for non-positive intervals, cross the right side of $\overline{CV}_1$ below 0.8 also. Hence $\overline{CV}_1 < \overline{T} < \overline{CV}_2$ and we do not reject H_0 which supports the hypothesis that $c = 0$. The same is true in the crisp case.

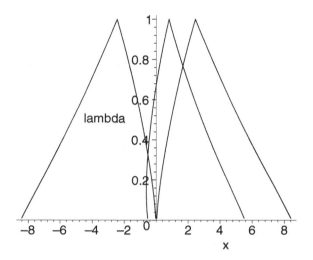

Figure 25.2: Fuzzy Test $\overline{T}$ verses the $\overline{CV}_i$ in Example 25.3.1 ($\overline{CV}_1$ left, $\overline{CV}_2$ right, $\overline{T}$ center)

25.4 References

1. J. Johnston: Econometric Methods, Second Edition, McGraw-Hill, N.Y., 1972.

2. Maple 9, Waterloo Maple Inc., Waterloo, Canada.

Chapter 26

Fuzzy One-Way ANOVA

26.1 Introduction

We first present the details of the non-fuzzy case. We follow the development in [1]. Then the fuzzy hypothesis test. We wish to compare the means of m populations where each population has the $N(\mu_i, \sigma^2)$ distribution with the same, but unknown, variance.

26.2 Crisp Hypothesis Test

The null hypothesis is

$$H_0 : \mu_1 = \mu_2 = ... = \mu_m = \mu, \tag{26.1}$$

with μ unknown and the alternative hypothesis H_1 is that the means are not all equal.

We collect independent random samples $X_{i1}, X_{i2}, ..., X_{i,n_i}$ from each population $1 \leq i \leq m$. Let $n = n_1 + ... + n_m$.

We introduce the "dot" notation used in ANOVA

$$\overline{X}_{\bullet\bullet} = \sum_{i=1}^{m} \sum_{j=1}^{n_i} X_{ij}/n, \tag{26.2}$$

and

$$\overline{X}_{i\bullet} = (1/n_i) \sum_{j=1}^{n_i} X_{ij}, \tag{26.3}$$

for $1 \leq i \leq m$, where $\overline{X}$ is not a fuzzy set but a mean. Next we define various sums of squares

$$SS(T) = \sum_{i=1}^{m} n_i (\overline{X}_{i\bullet} - \overline{X}_{\bullet\bullet})^2, \tag{26.4}$$

James J. Buckley: *Fuzzy Probability and Statistics*, StudFuzz **196**, 203–207 (2006)
www.springerlink.com   © Springer-Verlag Berlin Heidelberg 2006

$$SS(E) = \sum_{i=1}^{m} \sum_{j=1}^{n_i} (X_{ij} - \overline{X}_{i\bullet})^2, \tag{26.5}$$

$$SS(TO) = \sum_{i=1}^{m} \sum_{j=1}^{n_i} (X_{ij} - \overline{X}_{\bullet\bullet})^2, \tag{26.6}$$

where "SS" denotes "sum of squares". Then

$$SS(TO) = SS(E) + SS(T). \tag{26.7}$$

It can be shown that $SS(TO)/\sigma^2$ has a χ^2 distribution with $n-1$ degrees of freedom, $SS(E)/\sigma^2$ has a χ^2 distribution with $n-m$ degrees of freedom and $SS(T)/\sigma^2$ also has a χ^2 distribution with $m-1$ degrees of freedom. So

$$F(m-1, n-m) = [SS(T)/(\sigma^2\{m-1\})]/[SS(E)/(\sigma^2\{n-m\})], \tag{26.8}$$

has a F distribution, degrees of freedom $m-1$ and $n-m$, since the random variables in the numerator and denominator are independent. We will use the notation $F(s,t)$ for the F probability distribution with s degrees of freedom in the numerator and t degrees of freedom in the denominator. One can now argue, see [1], that if H_0 is true this $F(m-1, n-m)$ should be close to one but if H_0 is false $F(m-1, n-m)$ will become considerably greater than one.

Let $F_\gamma(m-1, n-m)$ be the value of $F(n-1, n-m)$ so that the probability of exceeding it equals γ. We compute this number from the F distribution with $m-1$ and $n-m$ degrees of freedom. This will be a one-sided test. Usually one uses $\gamma = 0.10, 0.05, 0.01$. So we reject H_0 if (from equation (26.8)) $F(m-1, n-m) \geq F_\gamma(m-1, n-m)$ and we do not reject H_0 otherwise.

Let $MS(T) = SS(T)/(m-1)$, $MS(E) = SS(E)/(n-m)$ and $MS(TO) = SS(TO)/(n-1)$ where "MS" denotes "mean square". Another fact that we will use in the next section is that under H_0 $MS(TO)$, $MS(T)$ and $MS(E)$ are all unbiased estimators of σ^2.

26.3   Fuzzy Hypothesis Test

We start with $MS(T)$ the numerator in equation (26.8), under H_0 it is an unbiased estimator of σ^2. Now $SS(T)/\sigma^2$ has a χ^2 distribution with $m-1$ degrees of freedom, so a $(1-\beta)100\%$ confidence interval for σ^2 would be (as in equation (9.4) in Chapter 9)

$$[(m-1)MS(T)/\chi^2_{R,\beta/2}, (m-1)MS(T)/\chi^2_{L,\beta/2}]. \tag{26.9}$$

However, this estimator is biased as shown in Chapter 9. Hence we define $L(\lambda)$ and $R(\lambda)$ as in equations (9.8) and (9.9), respectively, to obtain an unbiased estimator. For the rest of this chapter we will use $0.01 \leq \beta \leq 1$. Let

$$L_1(\lambda) = (1 - \lambda)\chi^2_{R,0.005} + \lambda(m - 1), \tag{26.10}$$

$$R_1(\lambda) = (1 - \lambda)\chi^2_{L,0.005} + \lambda(m - 1), \tag{26.11}$$

for $0 \leq \lambda \leq 1$. In the above equations the χ^2 distribution has $m - 1$ degrees of freedom. Then our fuzzy estimator of σ^2 based on $MS(T)$ has λ-cuts

$$[\frac{m-1}{L_1(\lambda)}MS(T), \frac{m-1}{R_1(\lambda)}MS(T)]. \tag{26.12}$$

Next we turn to $MS(E)$ in the denominator of equation (26.8), under H_0 it is also an unbiased estimator of σ^2. Now $SS(E)/\sigma^2$ has a χ^2 distribution with $n - m$ degrees of freedom, so a $(1 - \beta)100\%$ unbiased (we skip the biased case) confidence interval for σ^2 is

$$[\frac{n-m}{L_2(\lambda)}MS(E), \frac{n-m}{R_2(\lambda)}MS(E)], \tag{26.13}$$

where

$$L_2(\lambda) = (1 - \lambda)\chi^2_{R,0.005} + \lambda(n - m), \tag{26.14}$$

$$R_2(\lambda) = (1 - \lambda)\chi^2_{L,0.005} + \lambda(n - m), \tag{26.15}$$

for $0 \leq \lambda \leq 1$. In the above equations the χ^2 distribution has $n - m$ degrees of freedom.

Now we may compute our fuzzy statistic $\overline{F}$ for this test. $\overline{F}$ is a triangular shaped fuzzy number with our fuzzy estimator of σ^2 based on $MS(T)$ divided by our fuzzy estimator of σ^2 based on $MS(E)$. Using interval arithmetic, all intervals are positive, we get for λ-cuts

$$\overline{F}[\lambda] = [\Gamma_1(\lambda)\frac{MS(T)}{MS(E)}, \Gamma_2(\lambda)\frac{MS(T)}{MS(E)}], \tag{26.16}$$

where

$$\Gamma_1(\lambda) = \frac{(m-1)R_2(\lambda)}{(n-m)L_1(\lambda)}, \tag{26.17}$$

$$\Gamma_2(\lambda) = \frac{(m-1)L_2(\lambda)}{(n-m)R_1(\lambda)}, \tag{26.18}$$

for $0 \leq \lambda \leq 1$. The relationship between λ and β, in equations (26.12) and (26.13), can be determined as in equations (9.12) and (9.13) in Chapter 9.

Since our test statistic is fuzzy the critical values will also be fuzzy. There will be only one fuzzy critical value here because this is a one-sided test. Let $\overline{CV}_2$ go with the critical value F_γ and define $\overline{CV}_2[\lambda] = [cv_{21}(\lambda), cv_{22}(\lambda)]$. We use λ here since $\overline{F}[\lambda]$ is now a function of $\lambda \in [0, 1]$. We show how to get

$cv_{22}(\lambda)$. The end points of a λ-cut of $\overline{CV}_2$ are computed from the end points of the corresponding λ-cut of $\overline{F}$. We see that to find $cv_{22}(\lambda)$ we solve

$$P(\Gamma_2(\lambda)\frac{MS(T)}{MS(E)} \geq cv_{22}(\lambda)) = \gamma, \tag{26.19}$$

for $cv_{22}(\lambda)$. The above equation is the same as

$$P(\frac{MS(T)}{MS(E)} \geq \frac{1}{\Gamma_2(\lambda)}cv_{22}(\lambda)) = \gamma. \tag{26.20}$$

But under H_0 $\frac{MS(T)}{MS(E)}$ has a F distribution so

$$cv_{22}(\lambda) = \Gamma_2(\lambda)F_\gamma(m-1, n-m). \tag{26.21}$$

By using the left end point of $\overline{F}[\lambda]$

$$cv_{21}(\lambda) = \Gamma_1(\lambda)F_\gamma(m-1, n-m). \tag{26.22}$$

Hence a λ-cut of $\overline{CV}_2$ is

$$[\Gamma_1(\lambda)F_\gamma(m-1, n-m), \Gamma_2(\lambda)F_\gamma(m-1, n-m)]. \tag{26.23}$$

In the above equation for $\overline{CV}_2[\lambda]$, γ is fixed, and λ ranges in the interval $[0,1]$. $\overline{CV}_2$ will be a triangular shaped fuzzy number.

Example 26.3.1

The data in this example comes from Problem 8.6-2 in [1]. This is a classical agricultural study. There are $i = 1, 2, 3, 4$ varieties of corn and let μ_i be the average yield in bushels per acre of type i. We will test at the $\gamma = 5\%$ significance level that $\mu_1 = \mu_2 = \mu_3 = \mu_4$. Four test plots for each of the four varieties of corn were planted. The yields in bushels per acre of the four varieties of corn are given in Table 26.1.

We see that $n = 16$, $m = 4$, $n - m = 12$ and $MS(T) = 129.43$, $MS(E) = 26.37$. The crisp $F = 129.43/26.37 = 4.91 > 3.49 = F_{0.05}(3, 12)$ and we reject H_0.

Now we proceed to the fuzzy test. We compute from equation (26.16) λ-cuts of our fuzzy test statistic $\overline{F}$ to be

$$[1.2270\frac{(1-\lambda)4.404 + 12\lambda}{(1-\lambda)9.348 + 3\lambda}, 1.2270\frac{(1-\lambda)23.34 + 12\lambda}{(1-\lambda)0.216 + 3\lambda}], \tag{26.24}$$

for $0 \leq \lambda \leq 1$. We put these λ-cuts together, one on top of another, to get the fuzzy test statistic $\overline{F}$. Next we compute form equation (26.23) the λ-cuts of the fuzzy critical value

$$[0.8725\frac{(1-\lambda)4.404 + 12\lambda}{(1-\lambda)9.348 + 3\lambda}, 0.8725\frac{(1-\lambda)23.34 + 12\lambda}{(1-\lambda)0.216 + 3\lambda}], \tag{26.25}$$

X_1	X_2	X_3	X_4
68.82	86.84	90.16	61.58
76.99	75.69	78.84	73.51
74.30	77.87	80.65	74.57
78.73	76.18	83.58	70.75

Table 26.1: Data in Example 26.3.1

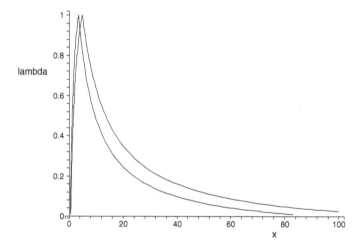

Figure 26.1: Fuzzy Test in Example 26.3.1($\overline{F}$ right, $\overline{CV}_2$ left)

for $\lambda \in [0,1]$. We put these λ-cuts together, one on top of another, to get the fuzzy critical value. The graphs of $\overline{F}$ and $\overline{CV}_2$, using Maple [2], are shown in Figure 26.1. We see that these two fuzzy numbers intersect above the level 0.80 so we decide $\overline{F} \approx \overline{CV}_2$ and we have no decision on H_0 in the fuzzy hypothesis test. Due to the uncertainty in the data in the fuzzy case we obtain "no decision" but in the crisp situation we got to reject H_0. The Maple commands for Figure 26.1 are in Chapter 30.

26.4   References

1. R.V. Hogg and E.A. Tanis: Probability and Statistical Inference, Sixth Edition, Prentice Hall, Upper Saddle River, N.J., 2001.

2. Maple 9, Waterloo Maple Inc., Waterloo, Canada.

Chapter 27

Fuzzy Two-Way ANOVA

27.1 Introduction

As in previous chapters we first review the non-fuzzy hypothesis tests and
then introduce the fuzzy hypothesis tests. This is also called the two-factor
analysis of variance. We will have two factors which we call factor A and
factor B. Assume factor A has a levels and factor B has b levels. There are
then $n = ab$ possible combinations each of which determines a cell. Let the
cells be arranged in a rows and b columns. In this chapter we have only one
observation per cell and we let X_{ij} denote the observation in the ith row and
j^{th} column. We assume that X_{ij} has a normal distribution $N(\mu_{ij}, \sigma^2)$, all
having a common variance, and the ab random variables are independent.

We will be able to test for a row effect and a column effect but not
for any interaction. You will need multiple observations per cell to test for
interaction. It is not difficult to extend the results in this chapter to handle
interactions. The development here follows that in [1].

27.2 Crisp Hypothesis Tests

We assume that the means μ_{ij} contain a row effect, a column effect and some
overall effect as

$$\mu_{ij} = \mu + \tau_i + \theta_j, \tag{27.1}$$

where τ_i represents the row effect and θ_j is for the column effect. Without
loss of generality (see [1]) we may assume that the sum of the τ_i (θ_j) equals
zero. There will be two tests. The first is that there is no row effect

$$H_{AO} : \tau_1 = \tau_2 = ... = \tau_a = 0. \tag{27.2}$$

James J. Buckley: *Fuzzy Probability and Statistics*, StudFuzz **196**, 209–217 (2006)
www.springerlink.com   © Springer-Verlag Berlin Heidelberg 2006

The alternate hypothesis is that they are not all equal to zero. The other test is that there is no column effect

$$H_{BO} : \theta_1 = \theta_2 = ... = \theta_b = 0. \tag{27.3}$$

The alternate hypothesis is that all the θ_i are not zero.

Now we have to define some notation and "sums of squares". Using the "dot" notation from ANOVA let

$$\overline{X}_{\bullet\bullet} = \sum_{i=1}^{a}\sum_{j=1}^{b} X_{ij}/ab, \tag{27.4}$$

$$\overline{X}_{i\bullet} = (1/b)\sum_{j=1}^{b} X_{ij}, \tag{27.5}$$

and

$$\overline{X}_{\bullet j} = (1/a)\sum_{i=1}^{a} X_{ij}, \tag{27.6}$$

where $\overline{X}$ is not a fuzzy set but a mean. Next we define various sums of squares

$$SS(A) = \sum_{i=1}^{a} b(\overline{X}_{i\bullet} - \overline{X}_{\bullet\bullet})^2, \tag{27.7}$$

$$SS(B) = \sum_{j=1}^{b} a(X_{\bullet j} - \overline{X}_{\bullet\bullet})^2, \tag{27.8}$$

$$SS(E) = \sum_{i=1}^{a}\sum_{j=1}^{b} (X_{ij} - X_{i\bullet} - X_{\bullet j} + X_{\bullet\bullet})^2, \tag{27.9}$$

$$SS(TO) = \sum_{i=1}^{a}\sum_{j=1}^{b} (X_{ij} - \overline{X}_{\bullet\bullet})^2, \tag{27.10}$$

where "SS" denotes "sum of squares". Then

$$SS(TO) = SS(A) + SS(B) + SS(E). \tag{27.11}$$

First assume that H_{AO} is true. Then it can be shown that $SS(A)/\sigma^2$ has a χ^2 distribution with $a-1$ degrees of freedom and $SS(A)/(a-1)$ is an unbiased estimator of σ^2 and $SS(E)/\sigma^2$ has a χ^2 distribution with $(a-1)(b-1)$ degrees of freedom and $SS(E)/(a-1)(b-1)$ is an unbiased estimator of σ^2. So

$$F_A(a-1, (a-1)(b-1)) = [SS(A)/(\sigma^2\{a-1\})]/[SS(E)/(\sigma^2\{(a-1)(b-1)\})], \tag{27.12}$$

has a F distribution, degrees of freedom $a - 1$ and $(a - 1)(b - 1)$, since the random variables in the numerator and denominator are independent. We will use the notation $F(s, t)$ for the F probability distribution with s degrees of freedom in the numerator and t degrees of freedom in the denominator. One can now argue, see [1], that if H_{AO} is true this $F(a - 1, (a - 1)(b - 1))$ should be close to one but if H_{AO} is false $F(a - 1, (a - 1)(b - 1))$ will become considerably greater than one.

Let $F_\gamma(a - 1, (a - 1)(b - 1))$ be the value of $F(a - 1, (a - 1)(b - 1))$ so that the probability of exceeding it equals γ. We compute this number from the F distribution with $a - 1$ and $(a - 1)(b - 1)$ degrees of freedom. This will be a one-sided test. Usually one uses $\gamma = 0.10, 0.05, 0.01$. So we reject H_{AO} if (from equation (27.12)) $F(a - 1, (a - 1)(b - 1)) \geq F_\gamma(a - 1, (a - 1)(b - 1))$ and we do not reject H_0 otherwise.

Next assume that H_{BO} is true. Then it can be shown that $SS(B)/\sigma^2$ has a χ^2 distribution with $b - 1$ degrees of freedom and $SS(B)/(b - 1)$ is an unbiased estimator of σ^2 and $SS(E)/\sigma^2$ has a χ^2 distribution with $(a-1)(b-1)$ degrees of freedom and $SS(E)/(a - 1)(b - 1)$ is an unbiased estimator of σ^2. So

$$F_B(b-1, (a-1)(b-1)) = [SS(B)/(\sigma^2\{b-1\})]/[SS(E)/(\sigma^2\{(a-1)(b-1)\})], \tag{27.13}$$

has a F distribution, degrees of freedom $b - 1$ and $(a - 1)(b - 1)$, since the random variables in the numerator and denominator are independent. One can now argue, see [1], that if H_{BO} is true this $F(b - 1, (a - 1)(b - 1))$ should be close to one but if H_{BO} is false $F(b - 1, (a - 1)(b - 1))$ will become considerably greater than one.

Let $F_\gamma(b - 1, (a - 1)(b - 1))$ be the value of $F(b - 1, (a - 1)(b - 1))$ so that the probability of exceeding it equals γ. We compute this number from the F distribution with $b - 1$ and $(a - 1)(b - 1)$ degrees of freedom. This will be a one-sided test. Usually one uses $\gamma = 0.10, 0.05, 0.01$. So we reject H_{BO} if (from equation (27.13)) $F(b - 1, (a - 1)(b - 1)) \geq F_\gamma(b - 1, (a - 1)(b - 1))$ and we do not reject H_0 otherwise.

Let $MS(A) = SS(A)/(a - 1)$, $MS(B) = SS(B)/(b - 1)$, $MS(E) = SS(E)/(a-1)(b-1)$ and $MS(TO) = SS(TO)/(ab-1)$ where "MS" denotes "mean square".

27.3   Fuzzy Hypothesis Tests

We first consider H_{AO}. Start with $MS(A)$ the numerator in equation (27.12), under H_{AO} it is an unbiased estimator of σ^2. Now $SS(A)/\sigma^2$ has a χ^2 distribution with $a - 1$ degrees of freedom, so a $(1 - \beta)100\%$ confidence interval for σ^2 would be (as in equation (9.4) in Chapter 9)

$$[(a - 1)MS(A)/\chi^2_{R,\beta/2}, (a - 1)MS(A)/\chi^2_{L,\beta/2}]. \tag{27.14}$$

However, this estimator is biased as shown in Chapter 9. Hence we define $L(\lambda)$ and $R(\lambda)$ as in equations (9.8) and (9.9), respectively, to obtain an unbiased estimator. For the rest of this chapter we will use $0.01 \le \beta \le 1$. Let

$$L_1(\lambda) = (1 - \lambda)\chi^2_{R,0.005} + \lambda(a - 1), \qquad (27.15)$$

$$R_1(\lambda) = (1 - \lambda)\chi^2_{L,0.005} + \lambda(a - 1), \qquad (27.16)$$

for $0 \le \lambda \le 1$. In the above equations the χ^2 distribution has $a - 1$ degrees of freedom. Then our fuzzy estimator of σ^2 based on $MS(A)$ has λ-cuts

$$[\frac{a-1}{L_1(\lambda)}MS(A), \frac{a-1}{R_1(\lambda)}MS(A)]. \qquad (27.17)$$

Next we turn to $MS(E)$ in the denominator of equation (27.12), under H_0 it is also an unbiased estimator of σ^2. Now $SS(E)/\sigma^2$ has a χ^2 distribution with $(a - 1)(b - 1)$ degrees of freedom, so a $(1 - \beta)100\%$ unbiased (we skip the biased case) confidence interval for σ^2 is

$$[\frac{(a - 1)(b - 1)}{L_2(\lambda)}MS(E), \frac{(a - 1)(b - 1)}{R_2(\lambda)}MS(E)], \qquad (27.18)$$

where

$$L_2(\lambda) = (1 - \lambda)\chi^2_{R,0.005} + \lambda(a - 1)(b - 1), \qquad (27.19)$$

$$R_2(\lambda) = (1 - \lambda)\chi^2_{L,0.005} + \lambda(a - 1)(b - 1), \qquad (27.20)$$

for $0 \le \lambda \le 1$. In the above equations the χ^2 distribution has $(a - 1)(b - 1)$ degrees of freedom.

Now we may compute our fuzzy statistic $\overline{F}_A$ for this test. $\overline{F}_A$ is a triangular shaped fuzzy number with our fuzzy estimator of σ^2 based on $MS(A)$ divided by our fuzzy estimator of σ^2 based on $MS(E)$. Using interval arithmetic, all intervals are positive, we get for λ-cuts

$$\overline{F}_A[\lambda] = [\Gamma_1(\lambda)\frac{MS(A)}{MS(E)}, \Gamma_2(\lambda)\frac{MS(A)}{MS(E)}], \qquad (27.21)$$

where

$$\Gamma_1(\lambda) = \frac{(a - 1)R_2(\lambda)}{(a - 1)(b - 1)L_1(\lambda)}, \qquad (27.22)$$

$$\Gamma_2(\lambda) = \frac{(a - 1)L_2(\lambda)}{(a - 1)(b - 1)R_1(\lambda)}, \qquad (27.23)$$

for $0 \le \lambda \le 1$. The relationship between λ and β, in equations (27.17) and (27.18), can be determined as in equations (9.12) and (9.13) in Chapter 9.

Since our test statistic is fuzzy the critical values will also be fuzzy. There will be only one fuzzy critical value here because this is a one-sided test. Let $\overline{CV}_2$ go with the critical value F_γ and define $\overline{CV}_2[\lambda] = [cv_{21}(\lambda), cv_{22}(\lambda)]$. We use λ here since $\overline{F}[\lambda]$ is now a function of $\lambda \in [0, 1]$. We show how to get

$cv_{22}(\lambda)$. The end points of a λ-cut of $\overline{CV}_2$ are computed from the end points of the corresponding λ-cut of $\overline{F}_A$. We see that to find $cv_{22}(\lambda)$ we solve

$$P(\Gamma_2(\lambda)\frac{MS(A)}{MS(E)} \geq cv_{22}(\lambda)) = \gamma, \qquad (27.24)$$

for $cv_{22}(\lambda)$. The above equation is the same as

$$P(\frac{MS(A)}{MS(E)} \geq \frac{1}{\Gamma_2(\lambda)}cv_{22}(\lambda)) = \gamma. \qquad (27.25)$$

But under H_0 $\frac{MS(A)}{MS(E)}$ has a F distribution so

$$cv_{22}(\lambda) = \Gamma_2(\lambda)F_\gamma(a-1,(a-1)(b-1)). \qquad (27.26)$$

By using the left end point of $\overline{F}_A[\lambda]$

$$cv_{21}(\lambda) = \Gamma_1(\lambda)F_\gamma(a-1,(a-1)(b-1)). \qquad (27.27)$$

Hence a λ-cut of $\overline{CV}_2$ is

$$[\Gamma_1(\lambda)F_\gamma(a-1,(a-1)(b-1)),\Gamma_2(\lambda)F_\gamma(a-1,(a-1)(b-1))]. \qquad (27.28)$$

In the above equation for $\overline{CV}_2[\lambda]$, γ is fixed, and λ ranges in the interval $[0,1]$. $\overline{CV}_2$ will be a triangular shaped fuzzy number.

Next consider H_{BO}. Start with $MS(B)$ the numerator in equation (27.13), under H_{BO} it is an unbiased estimator of σ^2. Now $SS(B)/\sigma^2$ has a χ^2 distribution with $b-1$ degrees of freedom, so a $(1-\beta)100\%$ confidence interval for σ^2 would be (as in equation (9.4) in Chapter 9)

$$[(b-1)MS(B)/\chi^2_{R,\beta/2},(b-1)MS(B)/\chi^2_{L,\beta/2}]. \qquad (27.29)$$

However, this estimator is biased as shown in Chapter 9. Hence we define $L(\lambda)$ and $R(\lambda)$ as in equations (9.8) and (9.9), respectively, to obtain an unbiased estimator. Let

$$L_3(\lambda) = (1-\lambda)\chi^2_{R,0.005} + \lambda(b-1), \qquad (27.30)$$

$$R_3(\lambda) = (1-\lambda)\chi^2_{L,0.005} + \lambda(b-1), \qquad (27.31)$$

for $0 \leq \lambda \leq 1$. In the above equations the χ^2 distribution has $b-1$ degrees of freedom. Then our fuzzy estimator of σ^2 based on $MS(B)$ has λ-cuts

$$[\frac{b-1}{L_3(\lambda)}MS(B),\frac{b-1}{R_3(\lambda)}MS(B)]. \qquad (27.32)$$

Next we turn to $MS(E)$ in the denominator of equation (27.13), under H_{BO} it is also an unbiased estimator of σ^2. Now $SS(E)/\sigma^2$ has a χ^2 distribution with $(a-1)(b-1)$ degrees of freedom, so a $(1-\beta)100\%$ unbiased (we skip the biased case) confidence interval for σ^2 is

$$[\frac{(a-1)(b-1)}{L_4(\lambda)}MS(E),\frac{(a-1)(b-1)}{R_4(\lambda)}MS(E)], \qquad (27.33)$$

where
$$L_4(\lambda) = (1 - \lambda)\chi^2_{R,0.005} + \lambda(a - 1)(b - 1), \qquad (27.34)$$

$$R_4(\lambda) = (1 - \lambda)\chi^2_{L,0.005} + \lambda(a - 1)(b - 1), \qquad (27.35)$$

for $0 \le \lambda \le 1$. In the above equations the χ^2 distribution has $(a - 1)(b - 1)$ degrees of freedom.

Now we may compute our fuzzy statistic $\overline{F}_B$ for this test. $\overline{F}_B$ is a triangular shaped fuzzy number with our fuzzy estimator of σ^2 based on $MS(B)$ divided by our fuzzy estimator of σ^2 based on $MS(E)$. Using interval arithmetic, all intervals are positive, we get for λ-cuts

$$\overline{F}_B[\lambda] = [\Gamma_3(\lambda)\frac{MS(B)}{MS(E)}, \Gamma_4(\lambda)\frac{MS(B)}{MS(E)}], \qquad (27.36)$$

where
$$\Gamma_3(\lambda) = \frac{(b - 1)R_4(\lambda)}{(a - 1)(b - 1)L_3(\lambda)}, \qquad (27.37)$$

$$\Gamma_4(\lambda) = \frac{(b - 1)L_4(\lambda)}{(a - 1)(b - 1)R_3(\lambda)}, \qquad (27.38)$$

for $0 \le \lambda \le 1$. The relationship between λ and β, in equations (27.32) and (27.33), can be determined as in equations (9.12) and (9.13) in Chapter 9.

Since our test statistic is fuzzy the critical values will also be fuzzy. There will be only one fuzzy critical value here because this is a one-sided test. Let $\overline{CV}_2$ go with the critical value F_γ and define $\overline{CV}_2[\lambda] = [cv_{21}(\lambda), cv_{22}(\lambda)]$. We use λ here since $\overline{F}[\lambda]$ is now a function of $\lambda \in [0, 1]$. We show how to get $cv_{22}(\lambda)$. The end points of a λ-cut of $\overline{CV}_2$ are computed from the end points of the corresponding λ-cut of $\overline{F}_B$. We see that to find $cv_{22}(\lambda)$ we solve

$$P(\Gamma_4(\lambda)\frac{MS(B)}{MS(E)} \ge cv_{22}(\lambda)) = \gamma, \qquad (27.39)$$

for $cv_{22}(\lambda)$. The above equation is the same as

$$P(\frac{MS(B)}{MS(E)} \ge \frac{1}{\Gamma_4(\lambda)}cv_{22}(\lambda)) = \gamma. \qquad (27.40)$$

But under H_{BO} $\frac{MS(B)}{MS(E)}$ has a F distribution so

$$cv_{22}(\lambda) = \Gamma_4(\lambda)F_\gamma(b - 1, (a - 1)(b - 1)). \qquad (27.41)$$

By using the left end point of $\overline{F}_B[\lambda]$

$$cv_{21}(\lambda) = \Gamma_3(\lambda)F_\gamma(b - 1, (a - 1)(b - 1)). \qquad (27.42)$$

Hence a λ-cut of $\overline{CV}_2$ is

$$[\Gamma_3(\lambda)F_\gamma(b - 1, (a - 1)(b - 1)), \Gamma_4(\lambda)F_\gamma(b - 1, (a - 1)(b - 1))]. \qquad (27.43)$$

In the above equation for $\overline{CV}_2[\lambda]$, γ is fixed, and λ ranges in the interval $[0, 1]$. $\overline{CV}_2$ will be a triangular shaped fuzzy number.

Car	Gas 1	Gas 2	Gas 3	Gas 4
1	16	18	21	21
2	14	15	18	17
3	15	15	18	16

Table 27.1: Data in Example 27.3.1

Example 27.3.1

The data in this example comes from Example 8.7-1 in [1]. Each of three cars was driven on the highway using each of four different brands of gasoline. The three cars all have automatic transmission, six cylinders, approximately the same weight and were driven at approximately the same maximum speed. The four brands of gasoline were all "regular". The miles per gallon for each of the 12 different combinations is shown in Table 27.1. We will test at the $\gamma = 5\%$ significance level for a row (A factor = car) effect and for a column (B factor = gas) effect.

First we test $H_{AO} : \tau_1 = ... = \tau_3 = 0$. We see that $a = 3$, $b = 4$ and we easily compute $SS(A) = 24$, $SS(B) = 30$ and $SS(E) = 4$. So $MS(A) = 12$, $MS(B) = 10$ and $MS(E) = 2/3$.

We next compute $\overline{F}_A$ and its corresponding $\overline{CV}_2$. We obtain from equation (27.21) that

$$\overline{F}_A[\lambda] = [6\frac{1.237(1 - \lambda) + 6\lambda}{7.378(1 - \lambda) + 2\lambda}, 6\frac{14.45(1 - \lambda) + 6\lambda}{0.051(1 - \lambda) + 2\lambda}], \qquad (27.44)$$

and we put these λ-cuts together, one on top of another, to get the fuzzy test statistic $\overline{F}_A$. Next we compute from equation (26.28)

$$\overline{CV}_2[\lambda] = [\frac{5.14}{3}\frac{1.237(1 - \lambda) + 6\lambda}{7.378(1 - \lambda) + 2\lambda}, \frac{5.14}{3}\frac{14.45(1 - \lambda) + 6\lambda}{0.051(1 - \lambda) + 2\lambda}], \qquad (27.45)$$

and we put these λ-cuts together, one on top of another, to get the fuzzy critical value. The graphs of $\overline{F}_A$ and $\overline{CV}_2$, using Maple [2], are shown in Figure 27.1. The Maple commands for this figure are in Chapter 30.

We see that these two fuzzy numbers intersect below the level 0.80 so we decide $\overline{F}_A > \overline{CV}_2$ and we reject H_{AO} in the fuzzy hypothesis test. The same is true in the crisp test.

Next we test $H_{BO} : \theta_1 = ... = \theta_4 = 0$. We compute $\overline{F}_B$ and its corresponding $\overline{CV}_2$. We obtain from equation (27.36) that

$$\overline{F}_B[\lambda] = [7.5\frac{1.237(1 - \lambda) + 6\lambda}{9.348(1 - \lambda) + 3\lambda}, 7.5\frac{14.45(1 - \lambda) + 6\lambda}{0.216(1 - \lambda) + 3\lambda}], \qquad (27.46)$$

and we put these λ-cuts together, one on top of another, to get the fuzzy test

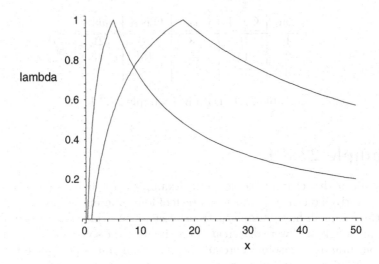

Figure 27.1: Fuzzy Test H_{AO} in Example 27.3.1($\overline{F}_A$ right, $\overline{CV}_2$ left)

statistic $\overline{F}_B$. Next we compute from equation (26.43)

$$\overline{CV}_2[\lambda] = [\frac{4.76}{2}\frac{1.237(1-\lambda)+6\lambda}{9.348(1-\lambda)+3\lambda}, \frac{4.76}{2}\frac{14.45(1-\lambda)+6\lambda}{0.216(1-\lambda)+3\lambda}], \qquad (27.47)$$

and we put these λ-cuts together, one on top of another, to get the fuzzy critical value. The graphs of $\overline{F}_B$ and $\overline{CV}_2$, using Maple [2], are shown in Figure 27.2.

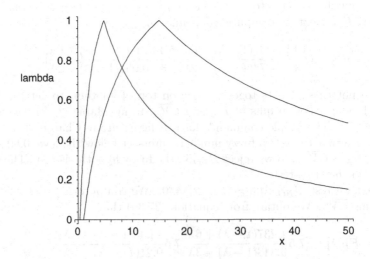

Figure 27.2: Fuzzy Test H_{BO} in Example 27.3.1($\overline{F}_B$ right, $\overline{CV}_2$ left)

We see that these two fuzzy numbers intersect below the level 0.80 so we decide $\overline{F}_B > \overline{CV}_2$ and we reject H_{BO} in the fuzzy hypothesis test. The same result holds for the crisp test.

27.4 References

1. R.V. Hogg and E.A. Tanis: Probability and Statistical Inference, Sixth Edition, Prentice Hall, Upper Saddle River, N.J., 2001.

2. Maple 9, Waterloo Maple Inc., Waterloo, Canada.

Chapter 28

Fuzzy Estimator
for the Median

28.1 Introduction

Estimating the median of a continuous probability distribution usually falls
into the area of non-parametric statistics, or distribution free statistics, and
comes at the end of a book on statistics. We first discuss a crisp method of
estimating the median. We follow the development in Section 7.7 in [1].

28.2 Crisp Estimator for the Median

Let X_i, $1 \le i \le n$, be a random sample from this continuous probability
distribution. If Y is a random variable having this continuous probability
distribution, then we define the median m to be

$$P(Y < m) = P(Y > m) = 0.50. \tag{28.1}$$

We are not assuming the continuous probability distribution is normal, or χ^2,
or exponential, etc. We order this random sample from smallest to largest
producing the order statistics $X_{(1)} < X_{(2)} < ... < X_{(n)}$. Here $X_{(1)}$ is the
smallest, $X_{(2)}$ is the next smallest, etc. We assume that no two values of the
X_i are equal. For simplicity assume n is odd. Let $\tau = (n+1)/2$. Then the
order statistic in the "middle" is $X_{(\tau)}$ and this will be our point estimate of
the median m. So $\widehat{m} = X_{(\tau)}$.

Next we need to construct confidence intervals for m. First we determine
the probability that the random interval $(X_{(1)}, X_{(n)})$ contains m. This is
not difficult because we have a binomial situation where the probability of a
success $p = 0.50$. Given an individual item X then $P(X < m) = 0.50$. In
order that $X_{(1)}$ to be less than m but $X_{(n)}$ to be greater than m we must

James J. Buckley: *Fuzzy Probability and Statistics*, StudFuzz **196**, 219–221 (2006)
www.springerlink.com

have at least one success but not n successes. So

$$P(X_{(1)} < m < X_{(n)}) = \sum_{i=1}^{n-1} \binom{n}{i} (0.50)^i (0.50)^{n-i} = 1 - \beta. \qquad (28.2)$$

Now let the values of the order statistics be $x_{(1)} < x_{(2)} < ... < x_{(n)}$. Then

$$(x_{(1)}, x_{(n)}), \qquad (28.3)$$

is a $(1 - \beta)100\%$ confidence interval for m. We generalize to

$$P(X_{(i)} < m < X_{(j)}) = \sum_{k=i}^{j-1} \binom{n}{k} (0.50)^k (0.50)^{n-k} = 1 - \beta. \qquad (28.4)$$

So

$$(x_{(i)}, x_{(j)}), \qquad (28.5)$$

is a $(1-\beta)100\%$ confidence interval for m. Equation (28.4) is used to determine the value of β.

Notice that this method only gives us a finite number of distinct confidence intervals for m.

28.3   Fuzzy Estimator

We place these confidence intervals, from the previous section, one on top of another, to build our fuzzy estimator $\overline{m}$ for m. However, this gives us only a finite number of confidence intervals to construct a triangular shaped fuzzy number $\overline{m}$. For example, with sample size $n = 11$ we have 5 confidence intervals. So we draw a smooth curve through the end points of these intervals to produce a triangular shaped fuzzy number $\overline{m}$. We illustrate this construction in the following example.

Example 28.3.1

Let $n = 9$ and assume the order statistics are

$$13.5 < 18.0 < 19.5 < 21.6 < 23.5 < 28.0 < 29.8 < 31.3 < 33.5. \qquad (28.6)$$

Them $\widehat{m} = 23.5$. We do the computations in equations (28.2)-(28.5) and the results are in Table 28.1.

Now we plot the points, which are shown in Figure 28.1, $(13.5, 0.004), (33.5, 0.004)$, $(18.0, 0.04), (31.3, 0.04)$, $(19.5, 0.18), ..., (23.5, 1)$ and then draw a smooth curve through these points producing $\overline{m}$. This is shown in Figure 28.1. This figure was done using LaTeX, not Maple. This is our triangular shaped fuzzy number estimator for the median.

Interval	β	Confidence
$(13.5, 33.5)$	0.004	99.6%
$(18.0, 31.3)$	0.04	96.0%
$(19.5, 29.8)$	0.18	82.0%
$(21.6, 28.0)$	0.508	49.2%
$(23.5, 23.5)$	1.0	0.0%

Table 28.1: Confidence Intervals in Example 28.3.1

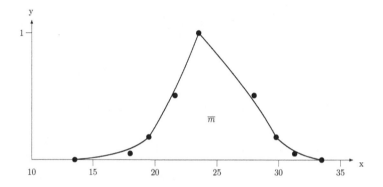

Figure 28.1: Fuzzy Estimator for the Median in Example 28.3.1

28.4 Reference

1. R.V. Hogg and E.A. Tanis: Probability and Statistical Inference, Sixth Edition, Prentice Hall, Upper Saddle River, N.J., 2001.

Chapter 29

Random Fuzzy Numbers

29.1 Introduction

We first, in this section, give a brief review of the literature on random fuzzy numbers. Then in the next section we present out method on generating a stream of random fuzzy numbers in some interval. Any random stream of numbers, fuzzy or crisp, must pass certain "randomness" tests and we perform one such test in Section 29.3. A Monte Carlo study, using our random fuzzy numbers, on a fully fuzzified linear program is discussed in the last section.

Mathematica has added random fuzzy numbers [25]. It can create "random" trapezoidal, Gaussian and triangular fuzzy numbers. They are represented by thin vertical bars similar to a histogram. We would need the functional expressions for the sides of the fuzzy numbers and it is not clear how we could get that information from Mathematica. The web site does not tell the user how these "random" fuzzy numbers are generated.

Most authors ([1],[12],[13],[18],[21],[22]) have used the following method to define random fuzzy numbers. Consider LR fuzzy numbers which we write as $(a, b, c)_{LR}$. Let m, l and r be three real-valued random variables with $m < l < r$. Then a random LR-fuzzy number is $(m, l, r)_{LR}$. The functions L and R are called the left and right shape functions, m is where the membership value equals one (vertex point) and $l(r) \geq 0$ is the left (right) spread of the fuzzy number. So, once you pick and fix L and R, the randomness is in the end points of the $\alpha = 0$ cut and the vertex point of the fuzzy number. We think the randomness should also be in the shape of the fuzzy number. That is, we should also be able to randomly change L and R.

Next, the paper [7] has another way to construct random fuzzy numbers. Let $F_i(x)$, $i = 1, 2, 3$, be probability distribution functions. Randomly choose $y \in [0, 1]$, then a random triangular fuzzy number has base $[F_1^{-1}(y), F_3^{-1}(y)]$ and vertex point at $F_2^{-1}(y)$. We have assumed that $F_1^{-1}(y) < F_2^{-1}(y) <$

James J. Buckley: *Fuzzy Probability and Statistics*, StudFuzz **196**, 223–234 (2006)
www.springerlink.com © Springer-Verlag Berlin Heidelberg 2006

$F_3^{-1}(y)$ for y in $[0,1]$. This just randomly produces a triangular fuzzy number but it always has the same shape (a triangle).

Finally, [20] generates a random triangular fuzzy number as $(m-6/m/m+2)$ for m uniform on $[1,3]$ and a random trapezoidal fuzzy number as $(m-4/m-2, m+4/m+6)$ for m a standard normal random variable. Again the shape is always the same, straight line segments for the sides of the fuzzy number.

Our random fuzzy numbers will have random base, random vertex point and also (limited) random shape. We believe this gives a better picture of random fuzzy numbers for fuzzy Monte Carlo methods.

29.2   Random Fuzzy Numbers

Let $\overline{N} \approx (n_1/n_2/n_3)$ in $[0,1]$ be a triangular shaped fuzzy number. In this section we will discuss how we plan to produce a sequence of random triangular shaped fuzzy numbers in some interval $[0, M]$, $M > 0$. We first make $\overline{N}$ in $[0,1]$ and multiply by M to get it in $[0, M]$.

Let $y = f_1(x)$ denote the function that makes the left side of the membership function $y = \overline{N}(x)$, $0 \leq y \leq 1$, $n_1 \leq x \leq n_2$. We assume that $f_1(x)$ is continuous and strictly increasing with $f_1(n_1) = 0$ and $f_1(n_2) = 1$. Next let $y = f_2(x)$ denote the function that makes the right side of the membership function $y = \overline{N}(x)$, $0 \leq y \leq 1$, $n_2 \leq x \leq n_3$. We assume that $f_2(x)$ is continuous and strictly decreasing with $f_2(n_2) = 1$ and $f_2(n_3) = 0$. Notice that if we substitute α for y an alpha-cut of $\overline{N}$ can be written $[f_1^{-1}(\alpha), f_2^{-1}(\alpha)]$.

In this chapter we will use quadratic functions for the $f_i(x)$. Let $a_{i1}x^2 + a_{i2}x + a_{i3} = f_i(x)$, $i = 1, 2$. We may easily extend the results to higher order polynomials. Now choose n_{11} and n_{21} so that $n_1 < n_{11} < n_2$, $n_2 < n_{21} < n_3$, and then choose y_1, y_2 in $(0, 1)$. The left side of $\overline{N}$ will be determined by the three points $(n_1, 0)$, (n_{11}, y_1), $(n_2, 1)$ because these three points, assuming they do not lie in a straight line, uniquely determine the a_{1j} in $y = a_{11}x^2 + a_{12}x + a_{13}$. The right side of $\overline{N}$ will be determined by the three points $(n_2, 1)$, (n_{21}, y_2), $(n_3, 0)$ because these three points, assuming they do not lie in a straight line, uniquely determine the a_{2j} in $y = a_{21}x^2 + a_{22}x + a_{23}$. So we require the seven numbers n_1, n_{11}, y_1, n_3, n_{21}, y_2 and n_3 to construct our triangular shaped fuzzy number $\overline{N}$. See Figure 29.1. We call these fuzzy numbers quadratic fuzzy numbers.

To randomly generate a $\overline{N}$ in $[0, 1]$ we randomly produce random numbers $x_1, ..., x_7$ in $[0, 1]$ giving the random vector $w = (x_1, ..., x_7) \in [0, 1]^7$. In w first randomly choose two values say, for example, x_3 and x_6. Then set $y_1 = x_3$ and $y_2 = x_6$. Now order the remaining five numbers from smallest to largest giving, for example, $x_5 < x_2 < x_7 < x_1 < x_4$. Then define $n_1 = x_5$, $n_{11} = x_2$, $n_2 = x_7$, $n_{21} = x_1$ and $n_3 = x_4$. We now have the five points to get triangular shaped fuzzy number $\overline{N}$. See Figure 29.1.

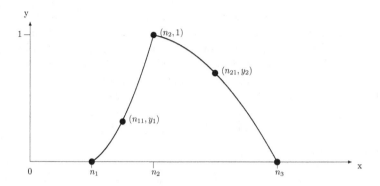

Figure 29.1: Random Quadratic Fuzzy Number $\overline{N}$

However, it is well known ([2],[10],[23])that these random vectors w will make clusters and empty regions in $[0,1]^7$. So we will use quasi-random numbers which will produce vectors w that uniformly fill the space $[0,1]^7$ ([2],[10],[23]).

How we propose to obtain a sequence of random triangular shaped fuzzy numbers is: (1) use quasi-random numbers to get random vectors w_1, w_2,...; and (2) each w_i makes a $\overline{N}_i$, $i = 1, 2, 3, ...$

29.3 Tests for Randomness

Assume we have a sequence of triangular shaped fuzzy numbers $\overline{N}_i$, $i = 1, 2, 3, ..., n$, in $[0, M]$ which we want to test for randomness. There are many randomness tests for sequences of real numbers ([9],[19]). But, most are not readily adapted to fuzzy numbers. The "run test" looks to be one of the easiest to apply to fuzzy numbers and we will use it in this section.

Make a new sequence using the symbols $+$, 0 and $-$ as follows: (1) if $\overline{N}_i < \overline{N}_{i+1}$ use $+$; (2) if $\overline{N}_i \approx \overline{N}_{i+1}$ use 0; and (3) if $\overline{N}_i > \overline{N}_{i+1}$ use $-$. From the original sequence of fuzzy numbers we get, for example

$$++0----+++++00------....0+++++++-----+++++++-------,$$
$$(29.1)$$

if $n = 56$. Now in the run test, applied to real number sequences, there will be no zeros. We could raise η to 0.9, or 0.95, see Section 2.5 in Chapter 2, but we will still get a few zeros. We will omit all the zeros and we obtain

$$++----+++++++-----+++...++++++-----+++++++------,$$
$$(29.2)$$

for $n = 50$. We count the total number of runs with the first run $++$, the second run $---$, third run $+++++$, etc. In our example above assume we get the total runs $s = 10$.

We do a statistical hypothesis test with null hypothesis H_0 the sequence of fuzzy numbers is random and the alternative hypothesis H_1 that the sequence is not random. We choose the level of significance (type I error) to be $\gamma = 0.05$. Under the null hypothesis the mean of s is $(2n-1)/3$ and the variance of s is $(16n-29)/90$. Also, we know that for large samples (say $n \geq 50$) the distribution of s is approximately normal ([9],[19]). In our example we have $n = 50$, the mean of s is 33 and the variance equals 8.5667. Doing a two sided test, and incorporating a continuity correction of 0.5, the left critical region is

$$(s + 0.5 - 33)/\sqrt{8.5667} \leq -1.96, \tag{29.3}$$

and the right critical region is

$$(s - 0.5 - 33)/\sqrt{8.5667} \geq 1.96. \tag{29.4}$$

So we reject H_0 when $s \leq 26$ or $s \geq 40$. In our example with $s = 10$ we reject H_0 and conclude that this sequence of fuzzy numbers is not random.

The left critical value guards against trends and the right critical value guards against cycles. A trend would be a sequences of increasing, or decreasing, fuzzy numbers leading to too few runs and $s \leq 26$. Cycles would produce something like $+ + - - + + - - + + - - \ldots.$ and too many runs with $s \geq 40$.

There are two other variations on the run test that could be used. Let $\overline{M}$ be the median of the sequence $\overline{N}_i$, $i = 1, ..., n$. The fuzzy median of a finite sequence of fuzzy numbers would need to be defined. Make a new sequence using the symbols $+$, 0 and $-$ as follows: (1) if $\overline{N}_i < \overline{M}$ use $-$; (2) if $\overline{N}_i \approx \overline{M}$ use 0; and (3) if $\overline{N}_i > \overline{M}$ use $+$. Omit the zeros. Count the runs below the median, count the runs above the median and let s be the total number of runs. Then using a normal approximation, similar to equations (29.3) and (29.4), we can do the hypothesis test of H_0 it is a random sequence versus H_1 it is not random ([11],[19]). A third test involves using the first sequence of $+'s$ and $-'s$ described above, not constructed from the median, and counting the length of a run of $+'s$ or a run of $-'s$. An asymptotically chi-square distributed test statistic based on the number of runs of length $L = 1, 2, 3, 4, 5$ and $L \geq 6$ is given in [17]. However, the author in [17] suggests a sample size of $n \geq 4000$ for a good approximation. But we shall use only the first run test given in equations (29.3) and (29.4) in this chapter.

Example 29.3.1

This example is the result of research on random fuzzy numbers done by Mr. Leonard J. Jowers [15] at the University of Alabama at Birmingham. We decided to implement this example in two steps. Both steps provide statistics on what they create. The first step was a crisp random number generator (RNGenerator) which prepares a file stream of crisp integer random numbers. The second step was RNAnalysis which takes a stream of crisp integer random numbers and generates random fuzzy numbers.

29.3.1 RNGenerator

The RNGenerator may be linked with any of several random number (RN) generator subroutines. The ones which we used were:

1. TrueRandom: A million 8-bit (in binary notation) true random numbers were downloaded from http://www.random.org. This routine supplies one 16-bit true random number (put two 8-bit numbers together), sequentially from that list, with each call.

2. PseudoRandom: This routine supplies one 16-bit pseudo random number from rand() with each call.

3. QuasiRandom: A Sobol quasi-random number routine from [24] was used as the basis for a quasi-random number generator (Section 7.7, "Quasi- Random Sequences," from [23] provides background to Sobol sequences). To make the use compatible with the other random number generators, it creates the Sobol numbers in vectors of length 5, but releases them one at a time with each call.

Type	Chi-Square	Minimum	Maximum	Equal Pairs
True	11.5182	0	32767	16
Pseudo	5.1051	0	32767	8
Quasi	219370	0	32766	59

Table 29.1: RNGenerator Chi-Square Results

RNGenerator does statistics on the stream of RNs it generates. A chi-square test is done for 12 bins (11 degrees of freedom) on 500,000 random numbers generated by each method. In Table 29.1: (1) "Chi-Square" is the value of the chi-square statistic; (2) "Minimum" is the smallest random number produced; (3) "Maximum" is the value of the largest random number generated; and (4) "Equal Pairs" means that two consecutively generated random numbers were equal.

The chi-square test was the standard randomness test applied to sequences of real numbers. The null hypothesis is H_0 that the sequence is random and the alternate hypothesis is H_1 that the sequence is not random. The significance level of the test was $\gamma = 0.05$. We place the random numbers into 12 equally spaced bins where, assuming H_0 is true, the expected number in each bin would be $500,000/12$. The critical value is $\chi^2 = 19.675$ for 11 degrees of freedom. So the true random numbers and the pseudo random numbers pass the randomness test (do not reject H_0) but the quasi-random numbers do not pass the test (reject H_0). The use of the quasi-random numbers will be explained in the next subsection.

29.3.2   RNAnalysis

Using streams of 500,000 "random" integers generated by the three methods, we completed the experiment in this second step. RNAnalysis creates quadratic fuzzy numbers (Section 29.2), displays them, and computes a randomness test on them.

In our preparation, we recognized that a quadratic fuzzy number could be represented by five points, one each at the left and right end points of the support, one at the vertex, and one each between the end points of the support and the vertex. We also realized that this representation in the computer requires a 7-tuple, a vector of length 7. See Figure 29.1. We also knew from [6] that random numbers suffer from a "curse of dimensionality", randomness quality degrades as the dimensionality of the vector increases. As explained in Section 29.2 the quasi-random method of producing vector w does not suffer from this problem. We will use the quasi-random procedure to get random vectors that will define our fuzzy numbers.

Next we are able to reduce the vector of length 7 to length 5 to obtain quadratic fuzzy numbers. This alternative representation, suggested by [16], is called Quadratic Bézier Generated Fuzzy Numbers (QBGFNs) and it requires only five numbers. QBGFNs are FNs whose membership functions, left and right, are defined by weighted quadratic rational Bézier curves ([8],[14]). This representation requires a vector of length 5. In addition to the three values normally required for a triangular fuzzy number, a left weight, and a right weight are needed. A full explanation of QBGFNs is lengthy and beyond the scope of this experiment, but is available from [15].

We now explain how we produce a quadratic fuzzy number from a "random" vector $v = (z_1, ..., z_5)$ generated from the quasi-random method. First we specify the range over which we allow the support, which we now assume to be $[0.0, 6.0]$, and the range for the two weights which we fix as $[-3.5, 3.5]$. The quasi-random method gives a vector of length 5 and we first map each component into $[0.0, 1.0]$. So assume $z_i \in [0, 1]$ all i. Let the final "random" vector, used for the quadratic fuzzy number $\overline{A}$ be $w = (x_1, ..., x_5)$. Then $x_3 = 6z_1$ gives the coordinate for the vertex of $\overline{A}$. If $\overline{A} \approx (a_1/a_2/a_3)$ then $a_2 = x_3 = z_1$, $x_1 = a_1 = x_3 - z_2 x_3$ and $x_5 = a_3 = x_3 + z_3(6 - x_3)$. Now we transform z_4 and z_5 into $[-3.5, 3.5]$ giving x_2 and x_4, respectively. The quadratic membership function on the left (right) is determined by x_2 (x_4). So we have $v = (z_1, ..., z_5)$ mapped to $w = (x_1, ..., x_5)$ translated to quadratic fuzzy number $\overline{A}$.

In Figure 29.2, which shows two QBGFNs, the example on the right is for the 5-tuple $(1.27, -3.50, 5.81, 1.53, 5.99)$.

As we create these QBGFNs, we apply the "run test" described in Section 29.3. Let the sequence of "random" fuzzy numbers be $\overline{A}_1, ..., \overline{A}_n$. The method used to decide on $\overline{A}_i < \overline{A}_{i+1}$, $\overline{A}_i \approx \overline{A}_{i+1}$ and $\overline{A}_i > \overline{A}_{i+1}$ is discussed in Section 2.5 using $\eta = 0.8$. If in the comparison of two fuzzy numbers in the sequence we get $\overline{A}_i \approx \overline{A}_{i+1}$, then we discard (reject) $\overline{A}_{i+1}$. Each test

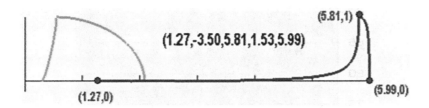

Figure 29.2: QBGFN Fuzzy Numbers

consists of evaluating QBGFNs until 100 are accepted. For a stream of 100 accepted FNs, as we showed in Section 29.3 for 50, the mean is 66.33334 and the variance is 17.45556. The null hypothesis (that the stream is random) is rejected, if the number of runs is ≤ 57 or ≥ 76 (similar to equations (29.3) and (29.4)).

Minimum	59
Maximum	77
Mean	67.76238
Standard Deviation	3.910101

Table 29.2: Summary Statistics on the Number of Runs for All Tests Using Quasi-Random Numbers

We replicated the test 100 times. We show, in Tables 29.2, 29.3 and Figure 29.3, results for the quasi-random number stream. On average, 16 QBGFNs were rejected in each test because of $\overline{A}_i \approx \overline{A}_{i+1}$. Notice from Figure 29.3 that the null hypothesis of randomness was only rejected twice in 100 tests. Using $\gamma = 0.05$ we would expect rejection, assuming randomness, on the average 5 times for 100 tests.

We also did this run test on sequences of quadratic fuzzy numbers generated from the true random method and from the pseudo random procedure. We summarize these results in Table 29.3. In Table 29.3 "Average Discards" is the average, over all 100 test runs, we had to discard a fuzzy number because $\overline{A}_i \approx \overline{A}_{i+1}$. Also, in Table 29.3 "Reject H_0" means the number of times, in 100 test runs, we rejected the null hypothesis of randomness.

Based on only one randomness test, the quasi-random method generated the "best" sequence of random fuzzy numbers. In the future we need to apply a few more randomness tests to our sequences of fuzzy numbers obtained by the quasi-random method. Assuming they also pass these new tests, then they will be used in our Monte Carlo studies.

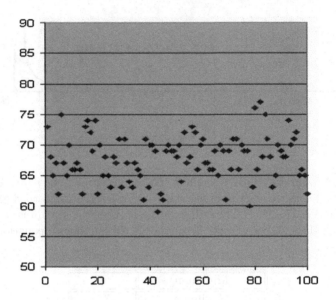

Figure 29.3: Results on the Number of Runs for All Tests Using Quasi-Random Numbers

Type	Mean	Minimum	Maximum	Average Discards	Reject H_0
True	70.7	61	84	41.6	16
Pseudo	71.5	61	78	44.1	15
Quasi	67.8	59	77	16.2	2

Table 29.3: Results of All "Runs Tests"

29.4   Monte Carlo Study

We consider one application of our fuzzy Monte Carlo method to the fully fuzzified linear program.

Fuzzy linear programming has long been an area of application of fuzzy sets. Consider the classical linear program

$$\max Z = c_1 x_1 + \cdots + c_n x_n$$

subject to: (29.5)

$$a_{i1} x_1 + \cdots + a_{in} x_n \leq b_i , \quad 1 \leq i \leq m$$

$$x_i \geq 0 \text{ , for all } i .$$

We need to have values for all the parameters c_i, a_{ij} and b_i to completely specify the optimization problem. Many of these must be estimated and are therefore uncertain. It is then natural to model these uncertain parameters using fuzzy numbers. The problem then becomes a fuzzy linear programming problem.

We are going to allow all the parameters to be fuzzy and we obtain what we have called the fully fuzzified linear programming problem. The fully fuzzified linear program is ([3]-[5])

$$max \overline{Z} = \overline{C}_1 \overline{X}_1 + \cdots + \overline{C}_n \overline{X}_n$$

$$\text{subject to:} \quad (29.6)$$

$$\overline{A}_{i1} \overline{X}_1 + \cdots + \overline{A}_{in} \overline{X}_n \leq \overline{B}_i, \ 1 \leq i \leq m,$$

$$\overline{X}_i \geq 0, \text{for all i}.$$

where the $\overline{C}_i$, $\overline{A}_{ij}$ and $\overline{B}_i$ can all be triangular fuzzy numbers. Since the parameters are fuzzy, the variables $\overline{X}_i$ will be triangular (shaped) fuzzy numbers. In the fuzzy Monte Carlo method the $\overline{X}_i$ are triangular shaped fuzzy numbers.

We will use the method in Section 2.5 of Chapter 2 for deciding if $\overline{M} \leq \overline{N}$ is true for two fuzzy numbers $\overline{M}$ and $\overline{N}$. We continue to employ $\eta = 0.8$ for the height of the intersection. Also, depending on the constraints, we need to decide on intervals $[0, I_i]$ for our random fuzzy numbers $\overline{X}_i$, $i = 1, 2, 3, ...$ We will do this in the example presented below. Also, for simplicity, assume that for this discussion $m = n = 2$. So we generate a stream of random fuzzy numbers and take them two at a time producing random fuzzy vectors $\overline{V} = (\overline{X}_1, \overline{X}_2)$ for $\overline{X}_i \in [0, I_i]$, $i = 1, 2$. Next we test $\overline{V}$ to see if it is feasible. $\overline{V}$ is feasible if

$$\overline{A}_{11} \overline{X}_1 + \overline{A}_{12} \overline{X}_2 \leq \overline{B}_1, \quad (29.7)$$

and

$$\overline{A}_{21} \overline{X}_1 + \overline{A}_{22} \overline{X}_2 \leq \overline{B}_2. \quad (29.8)$$

Once we have a feasible $\overline{V}$ we compute

$$\overline{Z} = \overline{C}_1 \overline{X}_1 + \overline{C}_2 \overline{X}_2. \quad (29.9)$$

Now go back to Section 2.5 of Chapter 2 and review the hierarchy of sets of fuzzy numbers $H_1, H_2, H_3, ... H_K$ where the highest ranked are all in H_K. It is a partitioning of all the $\overline{Z}$ values we get from feasible $\overline{V}$. Let the $\overline{Z}$ values from feasible $\overline{V}$ be $\overline{Z}_i$, $i = 1, 2, 3, ...$ Then if $\overline{Z}_a$ and $\overline{Z}_b$ are in H_K we get $\overline{Z}_a \approx \overline{Z}_b$ and if $\overline{Z}_c \in H_j$, $j < K$, then $\overline{Z}_a > \overline{Z}_c$. The general solution to the fuzzy optimization problem will be H_K. For each $\overline{Z} \in H_K$ we have stored the corresponding $\overline{V}$. If we require a single solution, show the $\overline{Z}$ in H_K to the decision maker(s) for them to pick one for the final solution.

However, if we produce $100,000$ random $\overline{V}$ we may end up with hundreds of $\overline{Z}$ values, with their corresponding $\overline{V}$, in H_K. Then we could employ another method to choose a "best" $\overline{Z}$ and $\overline{V}$ in H_K as used in ([3]-[5]).

Fuzzy Product Mix Problem

A company produces three products P_1, P_2 and P_3 each of which must be processed through three departments D_1, D_2 and D_3. The approximate time,

Department

		D_1	D_2	D_3
	P_1	6	12	2
Product	P_2	8	8	4
	P_3	3	6	1

Table 29.4: Approximate Times Product P_i is in Department D_j

in hours, each P_i spends in each D_j is given in Table 29.4.

Each department has only so much time available each week. These times can vary slightly from week to week so the following numbers are estimates of the maximum time available per week, in hours, for each department: (1) for D_1 288 hours; (2) 312 hours for D_2; and (3) D_3 has 124 hours. Finally, the selling price for each product can vary a little due to small discounts to certain customers but we have the following average selling prices: (1) $6 per unit for P_1; (2) $8 per unit for P_2, and (3) for P_3 $6/unit. The company wants to determine the number of units to produce for each product per week to maximize its revenue.

Since all the numbers given are uncertain, we will model the problem as a fully fuzzified linear program. We substitute a triangular fuzzy number for each value given where the peak of the fuzzy number is at the number given. So, we have the following fully fuzzified linear program

$$\max \overline{Z} = (5.8/6/6.2)\overline{X}_1 + (7.5/8/8.5)\overline{X}_2 + (5.6/6/6.4)\overline{X}_3 \qquad (29.10)$$

subject to:

$$(5.6/6/6.4)\overline{X}_1 + (7.5/8/8.5)\overline{X}_2 + (2.8/3/3.2)\overline{X}_3 \leq (283/288/293) ,$$
$$(11.4/12/12.6)\overline{X}_1 + (7.6/8/8.4)\overline{X}_2 + (5.7/6/6.3)\overline{X}_3 \leq (306/312/318) ,$$
$$(1.8/2/2.2)\overline{X}_1 + (3.8/4/4.2)\overline{X}_2 + (0.9/1/1.1)\overline{X}_3 \leq (121/124/127) ,$$
$$\overline{X}_1, \overline{X}_2, \overline{X}_3 \geq 0 ,$$

where the $\overline{X}_i$ are triangular (shaped) fuzzy numbers for the amount to produce for P_i per week.

Now we need to find the intervals $[0, I_i]$ for the $\overline{X}_i$, $i = 1, 2, 3$. This can be fairly complicated depending on the number of constraints and the method used to determine if $\overline{M} \leq \overline{N}$ is true. It is important to accurately determine these intervals because: (1) if they are too big the Monte Carlo simulation will be inefficient in producing too many infeasible fuzzy vectors; and (2) if they are too small we can miss a solution.

For comparison to the fuzzy solution we need to calculate the solution to the crisp linear program which is the one obtained by using the vertex values (where the membership function is one) of all the $\overline{C}_i$, $\overline{A}_{ij}$ and $\overline{B}_i$. The crisp solution is $x_1 = 0$ for (P_1), $x_2 = 27$ (for P_2) and $x_3 = 16$ (for P_3) with $\max z = 312$.

We will not continue solving this problem since it will take us further away from the basic topics of fuzzy probability and fuzzy statistics of this book. Solution will be contained in a future book on this topic.

29.5 References

1. H. Bandemer and A. Gebhardt: Bayesian Fuzzy Kriging, www-stat.uni-klu.ac.at/~agebhard/fuba/.

2. P. Bratley and B.L. Fox: Implementing Sobol's Quasi-Random Sequence Generator, ACM Trans. Math. Software, 14(1988)88-100.

3. J.J. Buckley and T. Feuring: Evolutionary Algorithm Solution to Fuzzy Problems: Fuzzy Linear Programming, Fuzzy Sets and Systems, 109(2000)35-53.

4. J.J. Buckley, E. Eslami and T. Feuring: Fuzzy Mathematics in Economics and Engineering, Physica-Verlag, Heidelberg, Germany, 2002.

5. J.J. Buckley, T. Feuring and Y. Hayashi: Multi-Objective Fully Fuzzified Linear Programming, Int. J. Uncertainty, Fuzziness and Knowledge-Based Systems, 9(2001)35-53.

6. David W. Deley: Computer Generated Random Numbers, http://world.std.com/~franl/crypto/random-numbers.html.

7. S. Ferson and L. Ginzburg: Hybrid Arithmetic, Proceedings ISUMA-NAFIPS, 1995, 619-623.

8. Gerald Farin: Curves and Surfaces for CADG, Fifth Edition, Academic Press, 2002, 219-224.

9. J.D. Gibbons and S. Chakraborti: Nonparametric Statistical Inference, Third Edition, Marcel Dekker, New York, 1992.

10. S.G. Henderson, B.A. Chiera and R.M. Cooke: Generating "Dependent" Quasi-Random Numbers, Proceedings of the 2000 Winter Simulation Conference, Orlando, FL, Dec.10-13, 2000, 527-536.

11. R.V. Hogg and E.A. Tanis: Probability and Statistical Inference, Sixth Edition, Prentice Hall, Upper Saddle River, NJ, 2001.

12. D.H. Hong and C.H. Ahn: Equivalent Conditions for Laws of Large Numbers for T-Related L-R Fuzzy Numbers, Fuzzy Sets and Systems, 136(2003)387-395.

13. M.D. Jimenez-Gamero, R. Pino-Mejias and M.A. Rojas-Medar: A Bootstrap Test for the Expectation of Fuzzy Random Variables, Comput. Statist. Data Anal.(2004).

14. J.K. Johnstone: Lecture Notes, CS-671, Computer Graphics and Modeling, Spring 2004, University of Alabama at Birmingham, Birmingham, Al., 35294.

15. L.J. Jowers: Department of Computer and Information Sciences, University of Alabama at Birmingham, Birmingham, Alabama, 35294, jowersl@cis.uab.edu.

16. L.J. Jowers: Fuzzy Numbers, A Novel Representation Using Bézier Curves, UAB Graduate Research Day 2005, March 4, 2005.

17. D.E. Knuth: The Art of Computer Programming, Vol. 2, Second Edition, Addison-Wesley, Reading, Mass., 1981.

18. R. Korner: An Asymptotic α−test for the Expectation of Random Fuzzy Variables, J. Statistical Planning and Inference, 83(2000)331-346.

19. B.W. Lindgren: Statistical Theory, Third Edition, Macmillan, New York, 1976.

20. Y.-K. Liu and Y.-J. Chen: Multicriteria Optimization Problem in Fuzzy Random Decision Systems, Proc. Third Int. Conf. Machine Learning and Cybernetics, Shanghai, China, 2004.

21. W. Nather: Linear Statistical Inference for Fuzzy Data, 3rd Int. Symposium Uncertainty Modeling and Analysis (ISUMA-NAFIPS), 1995, 71-XX.

22. W. Nather and R. Korner: Linear Regression with Random Fuzzy Numbers, in: B.M. Ayyub and M.M. Gupta (eds.), Uncertainty Analysis in Engineering and the Sciences, Kluwer, Boston, 1998.

23. W.H. Press, B.P. Flannery, S.A. Teukolsky, W.T. Vetterling: Numerical Recipes in C: The Art of Scientific Computing, Second Edition, Cambridge Univ. Press, 2002.

24. Sobol.c File Reference. `http://totil.udg.es/doc/sir/sobol_8C-source.html`.

25. www.wolfram.com/products/mathematica.

Chapter 30

Selected Maple/Solver Commands

30.1 Introduction

This chapter contains selected Maple and Solver commands for optimization problems and figures in this book. We start with Solver.

30.2 SOLVER

SOLVER is an optimization package which is an add-in to Microsoft Excel. It is free and if your Excel does not have it, then contact Microsoft or ([1],[3]). We used SOLVER because of the wide availability of Excel.

We will discuss how SOLVER was used to solve the four optimization problems in Examples 13.3.1 - 13.3.4. Excel is a spread sheet whose columns are labeled A,B,C,... and rows are labeled 1,2,3,... So "cell" B4 means the cell in the fourth row and B column. When we say $H2 = \mathcal{K}$ we mean put into cell H2 the formula/expression/data $\mathcal{K}$.

30.2.1 Example 13.3.1

1. $A1 = 0, A2 = 1, ..., A5 = 4$ (the x_i values)

2. $B1 = 0, B2 = 1, ..., B5 = 16$ (the x_i^2 values)

3. $C1 = p_1,...,C5 = p_5$ (initial values for the p_i)

4. $D1 = LN(C1), ..., D5 = LN(C5)$ (LN=natural log)

5. $E1 = SUM(C1 : C5)$ (sum of the p_i)

6. $F1 = SUMPRODUCT(A1 : A5 * C1 : C5)$ (mean=$M(p)$)

James J. Buckley: *Fuzzy Probability and Statistics*, StudFuzz **196**, 235–251 (2006)
www.springerlink.com © Springer-Verlag Berlin Heidelberg 2006

7. $G1 = SUMPRODUCT(B1 : B5 * C1 : C5)$

8. $G2 = G1 - (F1 * F1)$ (variance=$\sigma^2(p)$)

9. $H1 = SUMPRODUCT(C1 : C5 * D1 : D5)$

10. $H2 = -H1/LN(5)$ $(F(p)/ln(5) = \overline{G}(p))$

Now open the SOLVER window. Do the following: (1) target cell= $H2$ (what to max); (2) changing cells= $C1 : C5$ (the variables); and (3) constraints are $0.000001 \leq p_i \leq 1$ for all i and $E1 = 1$, $F1 = 3$. Choose "max" and click the solve button. A p_i can not be zero because of the $ln(p_i)$. In the options box we used automatic scaling, estimates=tangent, derivatives=forward, and search=Newton.

30.2.2   Example 13.3.2

We take SOLVER from Example 13.3.1 and add a few things for each subproblem. First consider subproblem #1. Add $I1 = F1 - 2$.

Open the SOLVER window. Then: (1) target cell= $J1$; (2) changing cells = $C1 : C5, J1$; (3) for the constraints change $F1 = 3$ to $2 \leq F1 \leq 3$ and add $J1 \leq H2$, $J1 \leq I1$. Choose max and click solve.

For subproblem #2 Make $I1 = 4 - F1$ and change $2 \leq F1 \leq 3$ to $3 \leq F1 \leq 4$.

30.2.3   Example 13.3.3

Assuming SOLVER was just used for Example 13.3.2 we do not need to change much for Example 13.3.3. Open the SOLVER window. Then: (1) target cell= $H2$; (2) changing cells $C1 : C5$; (3) with constraints changed as omit $2 \leq F1 \leq 3$, omit $J1 \leq H2$ and omit $J1 \leq I1$ and add $F1 = 2$, $G2 = 1$. Choose max and click solve.

30.2.4   Example 13.3.4

Assume that SOLVER just finished subproblem #1 Example 13.3.2.

1. Subproblem #1. Make $I1 = 3 - F1$ and $I2 = 2 - G2$. Open the SOLVER widow. Add constraints $1 \leq G2 \leq 2$, $J1 \leq I2$ Solve the max problem.

2. Subproblem #2. Change $I1$ to $I1 = F1 - 1$. Change the constraints to $1 \leq F1 \leq 2$. Solve.

3. Subproblem #3. Change $I2$ to $I2 = G2$. Change the constraints to $0 \leq G2 \leq 1$. Solve.

4. Subproblem #4. Change $I1$ to $I1 = 3 - F1$. Change the constraints to $2 \leq F1 \leq 2$. Solve.

30.2.5 Problems

The major problems with SOLVER are: (1) it can get out of the feasible set; and (2) it can stop at a local maximum. If Solver gives an error message (left the feasible set) just abort that run and start again. To guard against finding local maximums you will need to run SOLVER many times (the more the better) with different initial conditions. The initial p_i values need not sum to one but they must be positive.

30.3 Maple

In this section we give some of the Maple commands to solve optimization problems and to construct figures in the book. The commands are mostly in Maple 6 and a few are for Maple 9. The Maple 6 commands should be good for Maple 7, 8 and 9 [2].

When you do a figure in Maple and then export it to LaTeX you get two files. The first one is in LaTeX and the second is a "eps" file. In Maple 6 go into the "eps" file and then go to the " (backslash)drawborder true def" line and change it to "(backslash)drawborder false def" if you do not want the border. Also for the figures we will use y in place of α (or λ) for the variable in the vertical axis.

30.3.1 Chapter 3

Maple commands for Example 3.4.3.4 are:

1. with(Optimization);

2. obj:=x+y;

3. cnsts:=$[x <= 0.3, x >= 0.1, y >= 0.3, y <= 0.7, z >= 0.2, z <= 0.4, x + y + z = 1]$;

4. LPSolve(obj,cnsts,maximize);

5. LPSolve(obj,cnsts);
   (for minimization problem)

Maple commands for Example 3.4.3.7 are:

1. with(Optimization);

2. y:=$exp(-6 * x) - exp(-10 * x)$;
   (after integrating equation (3.30))

3. NLPSolve(y,x=2..6);
   (min. problem equation (3.32))

4. NLPSolve(y,x=2..6,maximize);
   (equation (3.33))

The Maple commands for Figure 3.6 are:

1. with(plots);

2. P111:=y->0.02+0.02*y;
   (using y for α)

3. P112:=y->0.06-0.02*y;

4. P212:=y->0.011-0.003*y;

5. P211:=y->0.005+0.003*y;

6. eq1:=x=P111(y)/(P111(y)+P212(y));

7. eq2:=x=P112(y)/(P112(y)+P211(y));

8. implicitplot({eq1,eq2},x=0..1,y=0..1,color=black,
   thickness=1,labels=[x,alpha]);

Maple commands for Figure 3.9 are:

1. with(plots);

2. N11:=y->0.06+0.02*y;
   (using y for α)

3. N12:=y->0.35+0.07*y;

4. D11:=y->0.55-0.05*y;

5. D12:=y->0.45+0.05*y;

6. N21:=y->0.10-0.02*y;

7. N22:=y->0.49-0.07*y;

8. D21:=y->0.45+0.05*y;

9. D22:=y->0.55-0.05*y;

10. eq1:=x=0.2*(N21(y)/D21(y))+0.7*(N22(y)/D22(y));

11. eq2:=x=0.2*(N11(y)/D11(y))+0.7*(N12(y)/D12(y));

12. implicitplot({eq1,eq2},x=0..1,y=0..1,color=black,
    thickness=1,labels=[x,alpha]);

30.3.2 Chapter 4

Maple commands for Figure 4.3 are:

1. with(plots);

2. f1:=y-> $1-(exp(-8-y)+(8+y)*exp(-8-y)+(8+y)^2*exp(-8-y)/2)$;

3. f2:=y-> $1 - (exp(-10 + y) + (10 - y) * exp(-10 + y) + (10 - y)^2 * exp(-10 + y)/2)$;

4. eq1:=x=f1(y);

5. eq2:=x=f2(y);

6. implicitplot({eq1,eq2},x=0..1,y=0..1,color=black,thickness=1,labels= [x,alpha]);

Maple commands for Figure 4.4 are:

1. with(plots);

2. with(stats);

3. f11:=y->statevalf[pf,binomiald[120,y]](115);

4. f12:=y->statevalf[pf,binomiald[120,y]](116)

5. f13:=y->statevalf[pf,binomiald[120,y]](117);

6. f14:=y->statevalf[pf,binomiald[120,y]](118);

7. f15:=y->statevalf[pf,binomiald[120,y]](119)

8. f16:=y->statevalf[pf,binomiald[120,y]](120);

9. f21:=y->statevalf[pf,binomiald[120,y]](115);

10. f22:=y->statevalf[pf,binomiald[120,y]](116);

11. f23:=y->statevalf[pf,binomiald[120,y]](117);

12. f24:=y->statevalf[pf,binomiald[120,y]](118);

13. f25:=y->statevalf[pf,binomiald[120,y]](119);

14. f26:=y->statevalf[pf,binomiald[120,y]](120);

15. f1:=y->f11(y)+f12(y)+f13(y)+f14(y)+f15(y)+f16(y);

16. f2:=y->f21(y)+f22(y)+f23(y)+f24(y)+f25(y)+f26(y);

17. eq1:=x=f1(0.75+0.10*y);

18. eq2:=x=f2(0.95-0.10*y);

19. implicitplot({eq1,eq2},x=0..1,y=0..1,color=black,thickness=1, labels=[x,alpha]);

30.3.3 Chapter 5

Maple commands for Example 5.3.1 are:

1. with(Optimization);

2. z:=int(0.398942*exp($-x^2/2$),x=(10-a)/b..(15-a)/b);

3. NLPSolve(z,a=8..12,b=2..$\sqrt{6}$);
   (gives the min for $\alpha = 0$)

4. NLPSolve(z,a=8..12,b=2..$\sqrt{6}$,maximize);
   (gives the max for $\alpha = 0$)

Maple commands for Example 5.5.3.1 are:

1. with(Optimization);

2. z:=int(0.398942*exp($-x^2/2$),x=(16.5-a)/$\sqrt{a}$..(21.5-a)/$\sqrt{a}$);

3. NLPSolve(z,a=15..25);
   (gives the min for $\alpha = 0$ cut for the normal)

4. NLPSolve(z,a=15..25,maximize);
   (gives the max for $\alpha = 0$ cut for the normal)

5. w:=sum($a^k exp(-a)/k!$,k=17..21);

6. NLPSolve(w,a=15..25);
   (gives the min for $\alpha = 0$ cut for the Poisson)

7. NLPSolve(w,a=15..25,maximize);
   (gives the max for $\alpha = 0$ cut for the Poisson)

30.3.4 Chapter 6

The Maple commands for Figure 6.2 are:

1. with(plots);

2. with(stats);

3. f1:=y->28.6-(1.25)*(statevalf[icdf,normald](1-(y/2)));

4. f2:=y->28.6+(1.25)*(statevalf[icdf,normald](1-(y/2)));

5. eq1:=x=f1(y);

6. eq2:=x=f2(y);

7. implicitplot({eq1,eq2},x=20..36,y=0.1..1,color=black,thickness=3,
   labels=[x,alpha]);
   (for the other graphs use $y = 0.01..1$ or $y = 0.001..1$)

30.3.5 Chapter 7

The Maple commands for Figure 7.1 are:

1. with(plots);

2. with(stats);

3. f1:=y->28.6-(0.3699)*(statevalf[icdf,studentst[24]](1-(y/2)));

4. f2:=y->28.6+(0.3699)*(statevalf[icdf,studentst[24]](1-(y/2)));

5. eq1:=x=f1(y);

6. eq2:=x=f2(y);

7. implicitplot({eq1,eq2},x=20..36,y=0.01..1,color=black,thickness=3,
   labels=[x,alpha]);
   (for the other graphs use $y = 0.10..1$ or $y = 0.001..1$)

30.3.6 Chapter 8

The Maple commands for Figure 8.1 are:

1. with(plots);

2. with(stats);

3. f1:=y->0.5143-(0.0267)*(statevalf[icdf,normald](1-(y/2)));

4. f2:=y->0.5143+(0.0267)*(statevalf[icdf,normald](1-(y/2)));

5. eq1:=x=f1(y);

6. eq2:=x=f2(y);

7. implicitplot({eq1,eq2},x=0..1,y=0.01..1,color=black,thickness=3,
   labels=[x,alpha]);
   (for the other graphs use $y = 0.10..1$ or $y = 0.001..1$)

30.3.7 Chapter 9

We substitute y for λ and then the Maple commands for Figure 9.1 are:

1. with(plots);

2. f1:=y->(1-y)*45.559 + y*24;

3. f2:=y->(1-y)*9.886 + y*24;

4. eq1:=x=82.08/f1(y);

5. eq2:=x=82.08/f2(y);

6. implicitplot({eq1,eq2},x=0..10,y=0.01..1,color=black,thickness=3,
   labels=[x,alpha]);
   (for the other graphs use $y = 0.10..1$ or $y = 0.001..1$)

30.3.8   Chapter 10

Maple commands for Figure 10.1 are:

1. with(plots);

2. with(stats);

3. f12:=y->statevalf[icdf,normald[0,1]](1-(y/2));

4. f11:=y-> $0.01 * (f12(y))^2 + 100$;

5. f1:=y->sqrt(f11(y));

6. f2:=y-> $(0.5 * f1(y) - 0.05 * f12(y))^2$;

7. eq1:=x=f2(y);

8. f3:=y-> $(0.5 * f1(y) + 0.05 * f12(y))^2$;

9. eq2:=x=f3(y);

10. implicitplot({eq1,eq2},x=20..30,y=0.01..1,color=black,
    thickness=1,labels=[x,alpha]);

Maple commands for Figure 10.2 are:

1. with(plots);

2. with(stats);

3. f1:=y->statevalf[icdf,normald[0,1]](1-0.5*y);

4. f2:=y->30/(20+f1(y));

5. f3:=y->30/(20-f1(y));

6. eq1:=x=f2(y);

7. eq2:=x=f3(y);

8. implicitplot({eq1,eq2},x=0..2,y=0.05..1,color=black,
   thickness=1,labels=[x,alpha]);

30.3.9   Chapters 11–13

No Maple commands are given.

30.3.10 Chapter 14

Maple commands for Example 14.4.1 are:

1. with(student);

2. m:=3;

3. f1:=x->(10*x*exp(10*x))/(exp(10*x)-1) - m*x -1;
   (Equations (14.13) and (14.14) together as one equation in only $x = \mu$)

4. u:=fsolve(f1(x),x);
   (solves for μ)

5. f2:=x->x/(exp(10*x)-1);
   (Equation (14.13)

6. c:=f2(u);

7. f3:=(-ln(c)-u*m)/ln(10);

8. evalf(%);
   (computes $\Omega(f)/ln(10)$)

Maple commands for Example 14.4.3 are:

1. with(student);

2. m:=3;

3. s:=1;
   (variance is one)

4. c:= $'c'$;

5. u:= $'u'$;
   (u=μ)

6. g:= $'g'$;
   (g=γ)

7. eq1:=$int(c * exp(u * x) * exp(g * (x - m)^2), x = 0..10) = 1$;

8. eq2:=$int(c * x * exp(u * x) * exp(g * (x - m)^2), x = 0..10) = m$;

9. eq3:=$int(c * (x - m)^2 * exp(u * x) * exp(g * (x - m)^2), x = 0..10)$=s;

10. sols:=fsolve({eq1,eq2,eq3},{c,u,g},{c=0..1,u=-1..0,g=-1..0});
    (Solutions are given)

Maple commands for Example 14.5.2 are:

1. with(student);

2. m:=3;

3. s:=1;
   (variance is one)

4. c:= $'c'$;

5. u:= $'u'$;
   (u=μ)

6. g:= $'g'$;
   (g=γ)

7. assume(u<0,g<0,$-u+6*g<0$);
   (so improper integrals converge)

8. eq1:=$int(c*exp(u*x)*exp(g*(x-m)^2),x=0..infinity)=1$;

9. eq2:=$int(c*x*exp(u*x)*exp(g*(x-m)^2),x=0..infinity)=m$;

10. eq3:=$int(c*(x-m)^2*exp(u*x)*exp(g*(x-m)^2),x=0..infinity)$=s;

11. sols:=fsolve({eq1,eq2,eq3},{c,u,g},{c=0..1,u=-1..0,g=-1..0});
    (Solutions are given)

30.3.11   Chapter 15

Maple commands for Figure 15.1 are:

1. with(plots);

2. with(stats);

3. f1:=y->1.6+(statevalf[icdf,normald](1-(y/2)));

4. f2:=y->1.6-(statevalf[icdf,normald](1-(y/2)));

5. f3:=y->1.96+(statevalf[icdf,normald](1-(y/2)));

6. f4:=y->1.96-(statevalf[icdf,normald](1-(y/2)));

7. eq1:=x=f1(y);

8. eq2:=x=f2(y);

9. eq3:=x=f3(y);

10. eq4:=x=f4(y);

11. implicitplot({eq1,eq2,eq3,eq4},x=-10..10,y=0.01..1,color=black,
    thickness=3,labels=[x,alpha]);

30.3.12   Chapter 16

We give the Maple commands for Figure 16.1. All the numerical values of
the data items have been loaded into the functions. We use y for λ.

1. with(plots);

2. with(stats);

3. f1:=y-> $\sqrt{1.40169 - 0.40169 * y}$;
   $(f_1(y) = \Pi_2)$

4. f2:=y-> $\sqrt{0.67328 + 0.32672 * y}$;
   $(f_2(y) = \Pi_1)$

5. L:=y-> $140.169 - 40.169 * y$;

6. R:=y-> $67.328 + 32.672 * y$;

7. z2:=1-statevalf[cdf,chisquare[100]](L(y));

8. z1:=statevalf[cdf,chisquare[100]](R(y));

9. z:=z1+z2;
   (computes $\alpha = z = f(\lambda)$, $y = \lambda$)

10. eq1:=x=f2(y)*(1.6-statevalf[icdf,studentst[100]](1-(z/2)));
    (the left side of $\overline{T}$)

11. eq2:=x=f1(y)*(1.6+statevalf[icdf,studentst[100]](1-(z/2)));
    (the right side of $\overline{T}$)

12. eq3:=x=f2(y)*(2.626-statevalf[icdf,studentst[100]](1-(z/2)));
    (the left side of $\overline{CV}_2$)

13. eq3:=x=f1(y)*(2.626+statevalf[icdf,studentst[100]](1-(z/2)));
    (the right side of $\overline{CV}_2$)

14. implicitplot({eq1,eq2,eq3,eq4},x=-10..10,y=0.01..1,color=black,
    thickness=3,labels=[x,alpha]);

30.3.13   Chapter 17

Maple commands for Figure 17.1 are:

1. with(plots);

2. with(stats);

3. f1:=y->0.8-(0.9968)*(statevalf[icdf,normald](1-(y/2)));

4. f2:=y->0.8+(0.9968)*(statevalf[icdf,normald](1-(y/2)));

5. f3:=y->1.96-(0.9968)*(statevalf[icdf,normald](1-(y/2)));

6. f4:=y->1.96-(0.9968)*(statevalf[icdf,normald](1-(y/2)));

7. eq1:=x=f1(y);

8. eq2:=x=f2(y);

9. eq3:=x=f3(y);

10. eq4:=x=f4(y);

11. implicitplot({eq1,eq2,eq3,eq4},x=-5..5,y=0.01..1,color=black, thickness=3,labels=[x,alpha]);

30.3.14 Chapter 18

The Maple commands for Figure 18.1 are, using y for λ:

1. with(plots);

2. L:=y->140.169 -40.169*y;

3. R:=y->67.328+32.672*y;

4. eq1:=x=8375/L(y);
   (left side $\overline{\chi}^2$)

5. eq2:=x=8375/R(y);
   (right side $\overline{\chi}^2$)

6. eq3:=x=6732.8/L(y);
   (left side $\overline{CV}_1$)

7. eq4:=x=6732.8/R(y);
   (right side $\overline{CV}_1$)

8. eq5:=x=100*140.169/L(y);
   (left side $\overline{CV}_2$)

9. eq6:=x=100*140.169/R(y);
   (right side $\overline{CV}_2$)

10. implicitplot({eq1,eq2,eq3,eq4,eq5,eq6 },x=0..300,y=0.01..1,color=black, thickness=3,labels=[x,alpha]);

30.3.15   Chapter 19

The Maple commands for Figure 19.1 are:

1. with(plots);

2. with(stats);

3. f1:=y->statevalf[icdf,normald](1-(y/2));

4. f2:=y->1.35-0.65*exp((2/sqrt(13))*f1(y));

5. f3:=y->1.35+0.65*exp((2/sqrt(13))*f1(y));

6. f4:=y->1.35-0.65*exp((-2/sqrt(13))*f1(y));

7. f5:=y->1.35+0.65*exp((-2/sqrt(13))*f1(y));

8. eq1:=x=f2(y)/f3(y);

9. eq2:=x=f4(y)/f5(y);

10. implicitplot({eq1,eq2},x=-1..1,y=0.01..1,color=black,thickness=3,
    labels=[x,alpha]);

30.3.16   Chapter 20

The Maple commands for Figures 20.1 and 20.2 are similar to those in Chapter 7 and are omitted.
The Maple commands for Figure 20.3 are (using y for λ):

1. with(plots);

2. f1:=y->(1-y)*21.955 + y*10;

3. f2:=y->(1-y)*1.344 + y*10;

4. eq1:=x=217.709/f1(y);

5. eq2:=x=217.709/f2(y);

6. implicitplot({eq1,eq2},x=0..50,y=0.01..1,color=black,thickness=3,
   labels=[x,alpha]);

30.3.17   Chapter 21

The Maple commands for Figure 21.1 are:

1. with(plots);

2. with(stats);

3. f1:=y->81.3-(1.6496)*(statevalf[icdf,studentst[8]](y/2));

4. f2:=y->81.3+(1.6496)*(statevalf[icdf,studentst[8]](y/2));
   (two sides of $\bar{a}$)

5. f3:=y->0.742+(0.1897)*(statevalf[icdf,studentst[8]](y/2));

6. f4:=y->0.742-(0.1897)*(statevalf[icdf,studentst[8]](y/2));
   (two sides of $\bar{b}$)

7. g1:=y-> f1(y)+(-8.3)*f4(y);

8. g2:=y-> f2(y)+(-8.3)*f3(y);

9. eq1:=x=g1(y);

10. eq2:=x=g2(y);

11. implicitplot({eq1,eq2},x=60..90,y=0.01..1,color=black,thickness=3,
    labels=[x,alpha]);

30.3.18   Chapter 22

The Maple commands for Figure 22.1 are (using y for λ):

1. with(plots);

2. with(stats);

3. f2:=y-> $\sqrt{2.1955 - 1.1955 * y}$;
   ($f_2(y) = \Pi_2$)

4. f1:=y-> $\sqrt{0.1344 + 0.8656 * y}$;
   ($f_1(y) = \Pi_1$)

5. L:=y-> $21.955 - 11.955 * y$;

6. R:=y-> $1.344 + 8.656 * y$;

7. z2:=1-statevalf[cdf,chisquare[8]](L(y));

8. z1:=statevalf[cdf,chisquare[8]](R(y));

9. z:=z1+z2;
   (computes $\alpha = z = f(\lambda)$, $y = \lambda$)

10. eq1:=x=f2(y)*(0.7880+statevalf[icdf,studentst[8]](1-(z/2)));
    (the right side of $\overline{T}$)

11. eq2:=x=f1(y)*(0.7880-statevalf[icdf,studentst[8]](1-(z/2)));
    (the left side of $\overline{T}$)

12. eq3:=x=f2(y)*(1.860+statevalf[icdf,studentst[8]](1-(z/2)));
    (the right side of $\overline{CV}_2$)

13. eq3:=x=f1(y)*(1.860-statevalf[icdf,studentst[8]](1-(z/2)));
    (the left side of $\overline{CV}_2$)

14. implicitplot({eq1,eq2,eq3,eq4},x=-10..10,y=0.01..1,color=black,
    thickness=3,labels=[x,alpha]);

30.3.19   Chapter 23

The Maple commands for Figures 23.1-23.3 are similar to those for Figure 7.1. The Maple commands for Figure 23.4 are like those for Figure 9.1.

30.3.20   Chapter 24

The Maple commands for Figure 24.1 are:

1. with(plots);

2. with(stats);

3. f11:=y->-49.3413-(24.0609)*(statevalf[icdf,studentst[6]](1-(y/2)));

4. f12:=y->-49.3413+(24.0609)*(statevalf[icdf,studentst[6]](1-(y/2)));

5. f21:=y->128*(1.3642-(0.1432)*(statevalf[icdf,studentst[6]](1-(y/2)));

6. f22:=y->128*(1.3642+(0.1432)*(statevalf[icdf,studentst[6]](1-(y/2)));

7. f31:=y->96*(0.1139-(0.1434)*(statevalf[icdf,studentst[6]](1-(y/2)));

8. f32:=y->96*(0.1139+(0.1434)*(statevalf[icdf,studentst[6]](1-(y/2)));

9. eq1:=x=f11(y)+f21(y)+f31(y);

10. eq2:=x=f12(y)+f22(y)+f32(y);

11. implicitplot({eq1,eq2},x=-100..400,y=0.01..1,color=black,thickness=3,
    labels=[x,alpha]);

30.3.21   Chapter 25

The Maple commands for Figures 25.1 and 25.2 are similar to those for Figure 22.1.

30.3.22   Chapter 26

The Maple commands for Figure 26.1 are (we use y for λ):

1. with(plots);

2. L1:=y->9.348-6.348*y;
   (using y for λ)

3. R1:=y->0.216+2.784*y;

4. L2:=y->23.34-11.34*y;

5. R2:=y-> 4.404+7.596*y;

6. eq1:=x=1.227*(R2(y)/L1(y));

7. eq2:=x=1.227*(L2(y)/R1(y));

8. eq3:=x=0.8725*(R2(y)/L1(y));

9. eq4:=x=0.8725*(L2(y)/R1(y));

10. implicitplot({eq1,eq2,eq3,eq4},x=1..100,y=0..1,color=black,
    thickness=1,labels=[x,lambda]);

30.3.23   Chapter 27

The Maple commands for Figure 27.1 are (we use y for λ):

1. with(plots);

2. L1:=y->7.378-5.378*y;
   (using y for λ)

3. R1:=y->0.051+1.949*y;

4. L2:=y->14.45-8.45*y;

5. R2:=y->1.237+4.763*y;

6. eq1:=x=6.0*(R2(y)/L1(y));

7. eq2:=x=6.0*(L2(y)/R1(y));

8. eq3:=x=1.713*(R2(y)/L1(y));

9. eq4:=x=1.713*(L2(y)/R1(y));

10. implicitplot({eq1,eq2,eq3,eq4},x=1..50,y=0..1,color=black, thickness=1,labels=[x,lambda]);

The Maple commands for Figure 27.2 are similar to those for Figure 17.1.

30.3.24 Chapter 28

The only figure was done in LaTeX.

30.3.25 Chapter 29

To obtain a copy of the C++ code to generate sequences of random fuzzy numbers described in Example 29.3.1 in Chapter 29 please contact Mr. Leonard J. Jowers, Department of Computer and Information Sciences, University of Alabama at Birmingham, Birmingham, Alabama, 35294, jowersl@cis.uab.edu.

30.4 References

1. Frontline Systems (www.frontsys.com).

2. Maple 9, Waterloo Maple Inc., Waterloo, Canada.

3. www.solver.com.

Chapter 31

Summary and Future Research

31.1 Summary

This book is an updated, and combination of, the two books [1] and [3]. Basically we took theoretical results from these two books and put them into this new book together with some new results.

First we took the end of Chapter 2 on restricted fuzzy arithmetic and Chapters 3, 4 and 8 of [1] into Chapters 3-5 of this book. We left out all the applications in [1]. What is new here is: (1) using a nonlinear optimization program in Maple [5] to solve certain optimization problems in fuzzy probability, where previously we used a graphical method or calculus; and (2) a new algorithm, suitable for using only pencil and paper, for solving some restricted fuzzy arithmetic problems. The introduction to fuzzy estimation in Chapters 6-9 is based on the book [3] and we refer the interested reader to that book for more about fuzzy estimators. The fuzzy estimators omitted from this book are those for $\mu_1 - \mu_2$, $p_1 - p_2$, σ_1/σ_2, etc. Fuzzy estimators for the arrival and service rates in Chapter 10 is from [2] and [4]. The reader should see those books for applications in queuing networks. Also, fuzzy estimators for the uniform probability density in Chapter 11 can be found in [4], but the derivation of these fuzzy estimators is new to this book. The fuzzy uniform distribution was used for arrival/service rates in queuing models in [4].

The estimation of fuzzy/crisp probability (mass) density functions based on a maximum entropy principle subject to fuzzy constraints in Chapters 12-14 are new to this book. In Chapter 12 we obtain fuzzy results but in Chapters 13 and 14 we determine crisp discrete and crisp continuous probability densities.

James J. Buckley: *Fuzzy Probability and Statistics*, StudFuzz **196**, 253–256 (2006)
www.springerlink.com © Springer-Verlag Berlin Heidelberg 2006

The introduction to fuzzy hypothesis testing in Chapters 15-18 is based on the book [3] and the reader needs to consult that book for more fuzzy hypothesis testing. What we omitted are tests on $\mu_1 = \mu_2$, $p_1 = p_2$, $\sigma_1 = \sigma_2$, etc.

The chapters on fuzzy correlation and regression come from [3]. The results on the fuzzy ANOVA (Chapters 26 and 27) and a fuzzy estimator for the median (Chapter 28) are new and have not been published before.

The chapter on random fuzzy numbers (Chapter 29) is also new to this book and these results have not been previously published. Applications of crisp random numbers to Monte Carlo studies are well known and we also plan to use random fuzzy numbers in Monte Carlo studies. Our first use of random fuzzy numbers will be to get approximate solutions to fuzzy optimization problems whose solution is unknown or computationally very difficult. However, this becomes a rather large project and will probably be the topic of a future book.

Chapter 30 contains selected Maple/Solver ([5]-[7]) commands used in the book to solve optimization problems or to generate the figures.

31.2   Future Research

Certain decisions were made in the book which will now be formulated as topics for future research.

31.2.1   Fuzzy Probability

More work can be done on the basic properties of our fuzzy probability including fuzzy conditional probability and fuzzy independence. There are other discrete, and continuous, fuzzy random variables to investigate together with their applications.

31.2.2   Unbiased Fuzzy Estimators

We faced the problem of getting an unbiased fuzzy estimator starting in Chapter 9. We said that our fuzzy estimator was unbiased if the vertex (where the membership function equals one) of the fuzzy number is at the crisp point estimator. For example, if we are estimating the variance of a normally distributed population, the vertex should be at s^2 the sample variance. Otherwise, it is a biased fuzzy estimator. Using the usual confidence intervals to construct our fuzzy estimator produced a biased fuzzy estimator. We proposed a solution to this problem giving an unbiased fuzzy estimator. Is there a "better" solution?

31.2.3 Comparing Fuzzy Numbers

In fuzzy hypothesis testing we need to be able to determine which of the following three possibilities for two fuzzy numbers $\overline{M}$ and $\overline{N}$ is true: (1) $\overline{M} < \overline{N}$, (2) $\overline{M} \approx \overline{N}$, or (3) $\overline{M} > \overline{N}$. In this book we used the procedure outlined in Section 2.5 in Chapter 2. If we use another method of comparing two fuzzy numbers (see the references in Chapter 2), how will the results of fuzzy hypothesis testing be effected? Is fuzzy hypothesis testing robust with respect to the method of comparing fuzzy numbers?

31.2.4 No Decision Conclusion

Starting in Chapter 15 our final conclusion in fuzzy hypothesis testing was: (1) reject the null hypothesis; (2) do not reject the null hypothesis; or (3) no decision on the null hypothesis. Let the fuzzy test statistic be $\overline{Z}$ and the two fuzzy critical values $\overline{CV}_i$, $i = 1, 2$. The "no decision" conclusion resulted from $\overline{CV}_1 \approx \overline{Z} < \overline{CV}_2$ or $\overline{CV}_1 < \overline{Z} \approx \overline{CV}_2$. In this case should the final decision be "do not reject the null hypothesis"?

31.2.5 Fuzzy Uniform

We need mathematical expressions for the confidence intervals in Chapter 11 so that we can accurately construct the fuzzy estimators.

31.2.6 Interval Arithmetic

These problems started in Chapter 16 where we were calculating $\alpha-$cuts of our fuzzy statistic from the quotient of two intervals $[a, b]/[c, d]$. We know that the interval in the denominator is usually positive ($c > 0$) but the interval in the numerator can be positive ($a > 0$), negative ($b < 0$) or "mixed" ($a < 0 < b$). Due to interval arithmetic (Section 2.3.2 in Chapter 2) we do the computation differently in the three cases of $[a, b]$ positive, negative or mixed. We discussed this in some detail in Section 16.3.1 in Chapter 16 but we did not implement these results formally in the rest of the book. For example, some of our graphs of our fuzzy statistic and its fuzzy critical values were not precisely correct. When this happened we mentioned this fact and that it did not effect the final result of reject H_0, do not reject H_0 or no decision on H_0. We need to correct this so that all the graphs are completely correct.

31.2.7 Fuzzy Prediction

We need theoretical results in Chapter 21 (and 24) on the comparison of $\overline{y}(x)[0]$, the 99% confidence interval for $E(Y)$ and the 99% confidence interval for y.

31.2.8   Fuzzy ANOVA

Extend the results in Chapter 27 to fuzzy two-way ANOVA with multiple data items per cell, and other models in this area.

31.2.9   Nonparametric Statistics

Nonparametric statistics is a large area and all we looked at in Chapter 28 was a fuzzy estimator for the median. Much more work can be done in this area of fuzzy nonparametric statistics.

31.2.10   Randomness Tests Fuzzy Numbers

We should expand our method of generating "random" fuzzy numbers to trapezoidal shaped fuzzy numbers. Also the sides of the fuzzy numbers may be described by polynomials of degree three or more. We need to develop more randomness tests for sequences of fuzzy numbers.

31.2.11   Future

In the Introduction we mentioned that we cover most of elementary statistics that can be found in an introductory course in statistic except contingency tables and nonparametric statistics. So where to next? Our method starts with crisp data producing fuzzy number estimators. Any statistical method based on estimation, not covered in this book, could be next.

31.3   References

1. J.J. Buckley: Fuzzy Probabilities: New Approach and Applications, Physica-Verlag, Heidelberg, Germany, 2003.

2. J.J. Buckley: Fuzzy Probabilities and Fuzzy Sets for Web Planning, Springer, Heidelberg, Germany, 2004.

3. J.J. Buckley: Fuzzy Statistics, Springer, Heidelberg, Germany, 2004.

4. J.J. Buckley: Simulating Fuzzy Systems, Springer, Heidelberg, Germany, 2005.

5. Frontline Systems (www.frontsys.com).

6. Maple 9, Waterloo Maple Inc., Waterloo, Canada.

7. www.solver.com.

Index

agricultural study, 206
algorithm
    new, 1, 29, 253
alpha
    function of λ, 91, 153, 155, 164, 173, 189
alpha-cut, 3, 4, 10, 13–16, 21, 26, 30–32, 34, 35, 41, 42, 45, 47, 52, 54, 56, 58, 60, 62–64, 67, 69, 72, 75, 76, 144, 152, 168, 177, 193, 198, 200, 224
approximations
    fuzzy binomial by fuzzy normal, 67
    fuzzy binomial by fuzzy Poisson, 57
    fuzzy Poisson by fuzzy normal, 70

Bayes' formula, 24, 40
    fuzzy, 40, 41, 47
beta
    function of λ, 91, 173, 189, 205, 212, 214
Bezier curves, 228
biased estimator
    factor, 90
binomial distribution, 3, 21, 85, 159, 219
bivariate normal distribution, 167
blood types, 41

calculus of variations, 110, 127, 130
chi-square distribution, 89, 90, 97, 145, 153, 163, 172, 188, 204, 210, 211, 219, 226

critical values, 3
    mean, 91
    median, 91
chi-square test
    randomness, 227
cockpit ejection seat design, 72
color blindness, 45
column effect, 209
confidence interval, 58, 72, 254
    $E(Y)$
        new value x, 179
        new values x, 195
    β, 3, 75
    λ, 96
    μ, 76, 77, 81, 82, 97
    ρ, 168
    σ, 91
    σ^2, 90, 92, 172, 173, 188, 204, 205, 211–213
    θ, 75
    a, 172, 173, 179, 187, 188, 195
    b, 172, 173, 179, 187, 188, 195
    c, 187, 188, 195
    p, 22, 85, 86
    y
        new value x, 179
        new values x , 195
    $y(x)$
        new value x , 179
    $y(x_1, x_2)$
        new values x, 195
    median, 219, 220
    uniform, 101
        details, 101
confidence region
    joint (a, b), 102

constraints
    linear, 26
continuity correction, 226
crisp
    constraints, 107, 115, 125, 129, 132
    function, 7
    matrix, 7
    probability, 24, 39, 40, 57
        discrete, 25
    solution, 8, 115
    subset, 7
crisp ANOVA
    dot notation, 203, 210
    SS notation, 210
crisp critical region, 181, 198, 226
crisp critical values, 144, 151, 204, 211
crisp data, 2, 4, 171, 187
crisp decision rule, 144, 151, 159, 163, 168, 181, 184, 198, 200, 204, 211, 226
    one-way ANOVA, 206
crisp estimator
    σ^2
        unbiased, 81, 204, 210, 211
    confidence interval, 76
    median, 219
    point, 76, 81, 254
        $\overline{x}$, 76
        ρ, 167
        σ^2, 171, 188
        θ, 75
        a, 171, 187
        a,b,c, 188
        b, 171, 187
        c, 187
        p, 21, 85, 159
        s^2, 89
        median, 219
        unbiased, 89
    uniform, 101
crisp hypothesis test
    μ

variance known, 143
variance unknown, 151
ρ, 167
σ^2, 163
a, 181
b, 183, 197
c, 199
p, 159
column effect, 210
one-way ANOVA, 203
randomness, 225, 256
row effect, 210
crisp linear regression, 171
crisp multiple regression, 187
crisp number, 3, 7
crisp set, 3, 7
crisp statistics, 2, 179, 195
    contingency tables, 2, 256
    distribution free, 219
    nonparametric, 2, 219, 256
crisp test statistic, 145
    μ
        variance known, 144
        variance unknown, 151
    ρ, 168
    σ^2, 163
    a, 181
    b, 184, 197
    c, 199
    p, 159
    column effect, 211
    one-way ANOVA, 204
    row effect, 210
curse of dimensionality, 228
cycles, 226

decision problem, 107–109, 111, 115, 116, 126
degrees of freedom, 3
dominated, 17

entropy, 108, 109, 116, 126
eps file, 237
Euler equation, 127, 130, 135, 137, 139

expert opinion, 24, 51, 52, 68, 107, 109–111, 115, 116, 126, 127
exponential distribution, 219
extension, 14, 15
extension principle, 3, 11, 13, 14, 16

F distribution, 145, 204, 211
 critical values, 3
failure, 21, 85
 probability, 21, 85
feasible set, 108, 110, 116, 126, 135, 232
figures, 4
 LaTeX, 4, 18, 35, 55, 60, 62, 64, 72, 104, 118, 129, 130, 220, 237
 Maple, 4, 5, 23, 45, 48, 56, 58, 59, 64, 70, 77, 82, 86, 92, 96, 98, 102, 147, 155, 160, 164, 168, 175, 178, 183, 194, 199, 207, 215, 216, 237
fuzzy
 constraints, 107, 115, 117, 125
fuzzy ANOVA, 1, 2, 254, 256
 one-way, 203
 two-way, 209
fuzzy arithmetic, 4, 11, 13, 36
 addition, 11
 alpha-cuts and interval arithmetic, 3, 144, 160, 164, 168, 177, 182, 184, 193, 198, 200, 205, 212, 214
 division, 11
 multiplication, 11, 39
 restricted, 4, 21, 24, 25, 31, 32, 52, 111
  linear programming, 32
 subtraction, 11
fuzzy arrival rate, 95
fuzzy correlation, 1, 167
fuzzy critical values, 2, 144, 152, 160, 164, 170, 182, 184, 198, 200, 205, 207, 212, 214
fuzzy data, 2, 4, 171, 187

fuzzy decision rule, 146, 149, 153, 155, 160, 164, 182, 184, 198
do not reject, 146, 156, 161, 164, 170, 183, 200, 255
no decision, 146, 147, 156, 165, 207, 255
reject, 146, 147, 162, 185, 199, 215, 217, 255
fuzzy estimator, 1, 4, 75, 111, 116, 127, 143
 $E(Y)$, 177
 $\overline{\lambda}$, 51, 59, 65, 71, 73, 96
 $\overline{\mu}$, 72, 75, 77, 82, 98, 144, 152
 $\overline{p}$, 168
 $\overline{\sigma}$, 72, 91, 94
 $\overline{\sigma}^2$, 89–92, 152, 164, 172, 175, 182, 184, 189, 190, 198, 200, 205, 212
 $\overline{a}$, 104, 172, 175, 177, 182, 188, 190, 193
 $\overline{b}$, 104, 172, 175, 177, 184, 188, 190, 193, 198
 $\overline{c}$, 188, 190, 200
 $\overline{m}$, 220
 $\overline{p}$, 22, 23, 51, 58, 68, 86, 160
 $\overline{y}(x_1, x_2)$, 193
 arrival rate, 1, 253
 biased, 89, 90, 172, 189, 204, 211, 213, 254
 median, 1, 2, 254, 256
 service rate, 1, 253
 unbiased, 89–91, 172, 189, 205, 212, 213, 254
 uniform, 2, 67, 101, 253, 255
fuzzy function, 13, 15, 16
 application, 15
fuzzy goal, 118, 119, 121, 128, 129, 132, 136, 137, 139
fuzzy hypothesis testing, 1, 2, 18, 254, 255
 μ
  variance known, 143
  variance unknown, 152
 ρ, 168

σ^2, 164
a, 181, 197
b, 181, 197
c, 197
p, 160
fuzzy independence, 38, 254
  properties, 39
  strong, 38, 39
  weak, 38, 40
fuzzy max entropy, 1, 107, 111, 112,
      115, 118, 121
  crisp density, 2, 253
  fuzzy density, 2, 253
fuzzy mean, 112, 113, 118, 121, 128,
      131, 135, 137, 139
fuzzy multiple regression, 187, 193
fuzzy nonparametric statistics, 256
fuzzy number, 8, 11–14, 21, 24, 25,
      30–32, 36, 38, 41, 43, 44, 52,
      75
  base, 10
  core, 10, 18, 38, 59
  discards, 229
  expert opinion, 23
  from confidence intervals, 75
  Gaussian, 223
  quadratic, 224, 226
  support, 10
  trapezoidal, 8, 59, 223
  trapezoidal shaped, 8, 60, 256
  triangular, 8, 37, 39, 40, 42, 43,
      45, 47, 48, 96, 98, 111, 116,
      127, 131, 136, 137, 223, 231
  triangular shaped, 4, 8, 22, 39,
      76, 77, 86, 144, 145, 153,
      160, 164, 172, 182, 184, 198,
      200, 205, 212, 214, 220, 224,
      231
fuzzy numbers, 1, 4, 7, 35, 177
  height of intersection, 18, 146,
      147, 155, 156, 160, 162, 164,
      170, 183, 185, 200, 207, 215,
      217
  partitioning, 17, 231

fuzzy optimization, 2, 118, 119, 121,
      128, 129, 132, 136, 137, 139,
      231, 254
fuzzy prediction, 177, 193, 255
fuzzy probability, 1, 9, 21, 24, 30–
      32, 39–41, 43, 52, 54, 55,
      57–59, 61–65, 67–69, 71–73,
      111, 112
  computing, 21, 26, 31
    calculus, 26–28
    feasible, 27, 34, 37, 42–44, 46
    first problem, 26
    graphical, 26, 28, 30
    Maple, 26, 28, 30, 31
    new algorithm, 29
    second problem, 30
    Solver, 26, 28
  conditional, 36, 38, 41, 44, 45,
      73, 254
    properties, 37
  confidence intervals, 21, 24
  discrete, 24, 25
    mean, 31
  expert opinion, 25
  generalized addition law, 33
  posterior, 24
  prior, 24
  properties, 33
fuzzy probability density
  uniform, 101
  joint, 112
  negative exponential, 31, 65, 73,
      113
    forgetfullness, 73
    fuzzy mean, 66
    fuzzy variance, 66
  normal, 30, 63, 69, 71, 72, 113
    fuzzy mean, 64
    fuzzy variance, 64
  uniform, 61, 67, 255
    fuzzy mean, 62
    fuzzy variance, 62
fuzzy probability distribution
  discrete

fuzzy mean, 34, 35
    fuzzy variance, 34, 35
    posterior, 41, 47
    prior, 47, 48
fuzzy probability mass function
    binomial, 51, 57, 58, 67, 69
      fuzzy mean, 52, 69
      fuzzy variance, 52, 69
    joint, 111
    marginals, 111
    Poisson, 51, 54, 55, 57, 71, 73
      fuzzy mean, 56
      fuzzy variance, 56
fuzzy queuing system, 95
fuzzy regression, 1, 171, 177
fuzzy service rate, 95, 97
fuzzy set, 1, 7
    discrete, 11
    subset, 3
fuzzy statistics, 4
fuzzy subset, 7, 11
fuzzy test statistic
    μ
      variance known, 144
      variance unknown, 152
    ρ, 168
    σ^2, 164
    a, 182
    b, 184, 198
    c, 200
    p, 160
    column effect, 214
    one-way ANOVA, 205, 206
    row effect, 212
fuzzy variance, 112, 113, 121, 131, 137, 139
fuzzy
    $\bar{c}$, 193

gamma distribution, 97
greatest lower bound, 12

imprecise side-conditions, 111, 116, 127
inequalities, 10

inf, 12
interval arithmetic, 4, 12–16, 25, 34, 177, 193, 255
    addition, 12
    division, 12
    multiplication, 12
    non-positive intervals, 153
    positive intervals, 152
    subtraction, 12

Lagrange multipliers, 108, 109, 127, 131, 135
least upper bound, 12
less than or equal, 11, 17, 46
linear correlation coefficient, 167
linear function, 171, 187
linear programming, 26, 230
    fully fuzzified, 230, 232
LR fuzzy numbers, 223

Maple, 92, 102, 117, 120, 134
    commands, 2, 5, 23, 59, 64, 70, 71, 78, 82, 86, 92, 96, 98, 128, 131, 137, 147, 155, 160, 175, 178, 183, 194, 207, 215, 237, 254
    commands Ex. 14.4.1, 243
    commands Ex. 14.4.3, 243
    commands Ex. 14.5.2, 243
    commands Ex. 3.4.3.4, 237
    commands Ex. 3.4.3.7, 237
    commands Ex. 5.3.1, 240
    commands Ex. 5.5.3.1, 240
    commands Fig. 10.1, 242
    commands Fig. 10.2, 242
    commands Fig. 15.1, 244
    commands Fig. 16.1, 245
    commands Fig. 17.1, 245
    commands Fig. 18.1, 246
    commands Fig. 19.1, 247
    commands Fig. 20.3, 247
    commands Fig. 21.1, 248
    commands Fig. 22.1, 248
    commands Fig. 24.1, 249
    commands Fig. 26.1, 250

commands Fig. 27.1, 250
commands Fig. 3.6, 238
commands Fig. 3.9, 238
commands Fig. 4.3, 239
commands Fig. 4.4, 239
commands Fig. 6.2, 240
commands Fig. 7.1, 241
commands Fig. 8.1, 241
commands Fig. 9.1, 241
implicitplot, 4, 78
optimization, 1, 5, 26, 32, 36,
    42, 63, 64, 72, 127, 128, 130,
    131, 137, 138, 140, 253
point estimators, 189
mathematica, 223
matrix
    X, 187
    identity, 188
    main diagonal, 188
matrix notation, 187
max entropy, 107, 108, 110, 115, 117,
    125, 126, 128, 131, 135, 137
maximum, 12
maximum likelihood estimator, 97
membership function, 7, 14
miles per gallon example, 215
minimum, 12
Monte Carlo, 1, 2, 232, 233, 254

normal distribution, 21, 76, 81, 82,
    85, 89, 92, 95, 97, 143, 145,
    151, 159, 163, 164, 168, 171,
    203, 209, 219, 226
    critical values, 3
notation
    fuzzy set, 3

one-sided test, 143, 149, 181, 197,
    204, 211
optimization problem, 26, 112, 230
    nonlinear, 35
order statistics, 219
ordering
    fuzzy numbers, 1, 3, 7, 17, 146,
        228, 231, 255

overbooking, 58

parameters, 230
polynomial functions, 256
prior information
    interval estimates, 108, 115
    mean, 108–110, 115, 116, 118,
        120, 121, 126–128, 135, 139
    variance, 108–110, 115, 116, 120,
        121, 126, 130, 136, 139
probability
    conditional, 40, 44, 47
    fuzzy, 4
    imprecise, 4
    interval, 4
probability density function, 75, 109,
    112, 125, 219
    negative exponential, 31, 65, 73,
        97, 110, 135
        forgetfullness, 73
    normal, 30, 63, 69, 70, 110, 113,
        139
    uncertain parameter, 30
    uniform, 61, 62, 67, 101, 127
probability distribution
    posterior, 40
    prior, 40
probability mass function, 75, 108,
    111, 115
    binomial, 51, 52, 57, 58, 68
    geometric, 109
    Poisson, 51, 54, 57, 59, 70, 73, 95
    uncertain parameter, 30
    uniform, 108, 116
product mix problem, 231
project scheduling, 23

QBGFNs, 228
quadratic functions, 224
quasi-random numbers, 225
queuing networks, 1, 253
queuing system, 101

random fuzzy numbers, 1, 2, 223, 224,
    226, 254, 256

code, 251
random fuzzy vector, 231
random numbers
    pseudo, 226
    quasi, 226
    true, 226
random sample, 2, 21, 24, 41, 43, 44,
        51, 75, 76, 81, 82, 85, 89, 92,
        95, 97, 101, 143, 167, 203,
        209, 219
    $\widehat{p}$, 159
    mean, 3, 76, 77, 81, 82, 144, 146,
        151, 155
    means, 204, 210
    variance, 81, 82, 151, 155, 163
random vector
    clusters, 225
randomness test, 229
ranking
    fuzzy numbers, 1, 3, 7, 18, 146,
        228, 231, 255
rapid response team, 59
resistance to surveys, 43
row effect, 209
run test, 225, 226, 229

sample correlation coefficient, 167
significance level
    γ, 3, 144, 146, 151, 159, 163, 168,
        181, 184, 197, 200, 226

significant linear correlation, 170
Solver, 119
    commands, 2, 5, 117, 121, 254
    commands Ex. 13.3.1, 235
    commands Ex. 13.3.2, 236
    commands Ex. 13.3.3, 236
    commands Ex. 13.3.4, 236
    optimization, 5, 117, 121
states of nature, 40, 47
subjective probability, 25
success, 21, 85
    probability, 21, 51, 85, 159, 219
sup, 12

t distribution, 81, 145, 151, 164, 172,
        181, 184, 188, 197, 200
    critical values, 3
testing HIV, 43
transitive, 17
trends, 226
two-sided test, 143

uncertainty, 9, 35, 44, 45, 59, 61, 72,
        108, 109, 116, 126, 171, 187
undominated, 17

vector, 187
    transpose, 187

List of Figures

2.1   Triangular Fuzzy Number $\overline{N}$ 8
2.2   Trapezoidal Fuzzy Number $\overline{M}$ 9
2.3   Triangular Shaped Fuzzy Number $\overline{P}$ 9
2.4   The Fuzzy Number $\overline{C} = \overline{A} \cdot \overline{B}$ 14
2.5   Determining $v(\overline{N} \leq \overline{M})$ 18

3.1   Fuzzy Estimator $\overline{p}$ in Example 3.2.1, $0.01 \leq \beta \leq 1$ 23
3.2   Fuzzy Variance in Example 3.5.2   35
3.3   Fuzzy Probability in the Blood Type Application   42
3.4   Fuzzy Probability in the Survey Application   44
3.5   Fuzzy Probability of HIV Given Test Positive   45
3.6   Fuzzy Probability of Male Given Color Blind   46
3.7   Fuzzy Probability of Female Given Color Blind   46
3.8 $\overline{P}(A_1)$ Using the Fuzzy Prior   48
3.9 $\overline{P}(A_1)$ Using the Fuzzy Posterior   48

4.1   Fuzzy Variance in Example 4.2.2   54
4.2   Fuzzy Probability in Example 4.3.1   55
4.3   Fuzzy Probability in Example 4.3.2   56
4.4   Fuzzy Probability of Overbooking   59
4.5   Fuzzy Probability of Multiple Attacks   60

5.1   Fuzzy Probability in Example 5.2.1   63
5.2   Fuzzy Probability in Example 5.3.1   65
5.3   Fuzzy Probability for the Fuzzy Exponential   66
5.4   Fuzzy Probability $\overline{P}[4,9]$ for the Fuzzy Uniform   68
5.5   Fuzzy Probability in the Ejection Seat Example   73

6.1   Fuzzy Estimator $\overline{\mu}$ in Example 6.3.1, $0.01 \leq \beta \leq 1$ 77
6.2   Fuzzy Estimator $\overline{\mu}$ in Example 6.3.1, $0.10 \leq \beta \leq 1$ 78
6.3   Fuzzy Estimator $\overline{\mu}$ in Example 6.3.1, $0.001 \leq \beta \leq 1$ 79

7.1   Fuzzy Estimator $\overline{\mu}$ in Example 7.1.1, $0.01 \leq \beta \leq 1$ 82

7.2 Fuzzy Estimator $\overline{\mu}$ in Example 7.1.1, $0.10 \leq \beta \leq 1$ 83
7.3 Fuzzy Estimator $\overline{\mu}$ in Example 7.1.1, $0.001 \leq \beta \leq 1$ 83

8.1 Fuzzy Estimator $\overline{p}$ in Example 8.1.1, $0.01 \leq \beta \leq 1$ 86
8.2 Fuzzy Estimator $\overline{p}$ in Example 8.1.1, $0.10 \leq \beta \leq 1$ 87
8.3 Fuzzy Estimator $\overline{p}$ in Example 8.1.1, $0.001 \leq \beta \leq 1$ 87

9.1 Fuzzy Estimator $\overline{\sigma}^2$ in Example 9.3.1, $0.01 \leq \beta \leq 1$ 92
9.2 Fuzzy Estimator $\overline{\sigma}^2$ in Example 9.3.1, $0.10 \leq \beta \leq 1$ 93
9.3 Fuzzy Estimator $\overline{\sigma}^2$ in Example 9.3.1, $0.001 \leq \beta \leq 1$ 93
9.4 Fuzzy Estimator $\overline{\sigma}$ in Example 9.3.1, $0.01 \leq \beta \leq 1$ 94

10.1 Fuzzy Arrival Rate $\overline{\lambda}$ in Example 10.2.1 97
10.2 Fuzzy Service Rate $\overline{\mu}$ in Example 10.3.1 98

11.1 Determining Joint Confidence Region for (a, b) 103
11.2 Fuzzy Estimators $\overline{a}$ and $\overline{b}$ in $U(a, b)$ 104
11.3 Fuzzy Estimators $\overline{a}$ and $\overline{b}$ in the Uniform Distribution 105

13.1 Fuzzy Goal $\overline{G}(p)$ for Example 13.3.2 118

14.1 Fuzzy Goal $\overline{G}(f)$ in Section 14.4 129
14.2 Solution to Example 14.4.2 130

15.1 Fuzzy Test $\overline{Z}$ verses $\overline{CV}_2$ in Example 15.3.1($\overline{Z}$ left, $\overline{CV}_2$ right) 147
15.2 Fuzzy Test $\overline{Z}$ verses $\overline{CV}_1$ in Example 15.3.1($\overline{CV}_1$ left, $\overline{Z}$ right) 148
15.3 Fuzzy Test $\overline{Z}$ verses $\overline{CV}_1$ in Example 15.3.2($\overline{Z}$ left, $\overline{CV}_1$ right) 148

16.1 Fuzzy Test $\overline{T}$ verses $\overline{CV}_2$ in Example 16.3.1($\overline{T}$ left, $\overline{CV}_2$ right) 155
16.2 Fuzzy Test $\overline{T}$ verses $\overline{CV}_1$ in Example 16.3.2($\overline{T}$ right, $\overline{CV}_1$ left) 156

17.1 Fuzzy Test $\overline{Z}$ verses $\overline{CV}_2$ in Example 17.3.1($\overline{Z}$ left, $\overline{CV}_2$ right) 161
17.2 Fuzzy Test $\overline{Z}$ verses $\overline{CV}_2$ in Example 17.3.2($\overline{Z}$ right, $\overline{CV}_2$ left) 161

18.1 Fuzzy Test in Example 18.3.1($\overline{CV}_1$ left, $\overline{\chi}^2$ middle, $\overline{CV}_2$ right) 165
18.2 Fuzzy Test in Example 18.3.2($\overline{CV}_1$ left, $\overline{\chi}^2$ middle, $\overline{CV}_2$ right) 165

19.1 Fuzzy Estimator $\overline{p}$ in Example 19.3.1 169
19.2 Fuzzy Test $\overline{Z}$ verses the $\overline{CV}_i$ in Example 19.3.2($\overline{CV}_1$ left, $\overline{Z}$ middle, $\overline{CV}_2$ right) 169

20.1 Fuzzy Estimator for a in Example 20.2.1 174
20.2 Fuzzy Estimator for b in Example 20.2.1 174
20.3 Fuzzy Estimator for σ^2 in Example 20.2.1 175

21.1 Fuzzy Estimator of $E(Y)$ given $x = 60$ in Example 21.1 . . . 178
21.2 Fuzzy Estimator of $E(Y)$ given $x = 70$ in Example 21.1 . . . 178

22.1 Fuzzy Test $\overline{T}$ verses $\overline{CV}_2$ in Example 22.2.1($\overline{T}$ left, $\overline{CV}_2$ right) 183
22.2 Fuzzy Test $\overline{T}$ verses the $\overline{CV}_i$ in Example 22.3.1 ($\overline{CV}_1$ left,
$\overline{CV}_2$ center, $\overline{T}$ right) 185

23.1 Fuzzy Estimator $\overline{a}$ for a in Example 23.2.1 190
23.2 Fuzzy Estimator $\overline{b}$ for b in Example 23.2.1 190
23.3 Fuzzy Estimator $\overline{c}$ for c in Example 23.2.1 191
23.4 Fuzzy Estimator $\overline{\sigma}^2$ for σ^2 in Example 23.2.1 191

24.1 Fuzzy Estimator of $E(Y)$ Given $x_1 = 128$, $x_2 = 96$, in Example
24.1.1 194
24.2 Fuzzy Estimator of $E(Y)$ Given $x_1 = 132$, $x_2 = 92$, in Example
24.1.1 194

25.1 Fuzzy Test $\overline{T}$ verses $\overline{CV}_2$ in Example 25.2.1($\overline{CV}_2$ left, $\overline{T}$ right) 199
25.2 Fuzzy Test $\overline{T}$ verses the $\overline{CV}_i$ in Example 25.3.1 ($\overline{CV}_1$ left,
$\overline{CV}_2$ right, $\overline{T}$ center) 201

26.1 Fuzzy Test in Example 26.3.1($\overline{F}$ right, $\overline{CV}_2$ left) 207

27.1 Fuzzy Test H_{AO} in Example 27.3.1($\overline{F}_A$ right, $\overline{CV}_2$ left) . . . 216
27.2 Fuzzy Test H_{BO} in Example 27.3.1($\overline{F}_B$ right, $\overline{CV}_2$ left) . . . 216

28.1 Fuzzy Estimator for the Median in Example 28.3.1 221

29.1 Random Quadratic Fuzzy Number $\overline{N}$ 225
29.2 QBGFN Fuzzy Numbers 229
29.3 Results on the Number of Runs for All Tests Using Quasi-
Random Numbers 230

List of Tables

3.1 Alpha-Cuts of $\overline{P}(O')$ 42

4.1 Fuzzy Poisson Approximation to Fuzzy Binomial 58

5.1 Alpha-Cuts of the Fuzzy Probability in Example 5.3.1 64
5.2 Fuzzy Normal Approximation to Fuzzy Binomial 70
5.3 Fuzzy Normal Approximation to Fuzzy Poisson 71
5.4 Alpha-Cuts of the $\overline{P}[140, 200]$ 72

9.1 Values of $factor$ for Various Values of n 90

11.1 Values of ϵ for Various Values of n, Given $\beta = 0.01$ 102
11.2 Values of ϵ for Various Values of n, Given $\beta = 0.50$ 103

13.1 Solution to Crisp Maximum Entropy in Example 13.3.1 . . . 117
13.2 Solution to Fuzzy Maximum Entropy in Example 13.3.2 . . . 120
13.3 Solution to Crisp Maximum Entropy in Example 13.3.3 . . . 121
13.4 Solution to Fuzzy Maximum Entropy in Example 13.3.4 . . . 123

14.1 Values of the Fuzzy Goal $\overline{G}(f^*)$ in Example 14.4.2 130
14.2 Values of the Fuzzy Goals in Example 14.4.4 134
14.3 Values of the Fuzzy Goals in Example 14.5.3 138

20.1 Crisp Data for Example 20.2.1 173

21.1 Comparing the 99% Confidence Intervals in Example 21.1 . . 179

23.1 Crisp Data for Example 23.2.1 189

24.1 Comparing the 99% Confidence Intervals in Example 24.1.1 . 195

26.1 Data in Example 26.3.1 207

27.1 Data in Example 27.3.1 215

28.1 Confidence Intervals in Example 28.3.1 221

29.1 RNGenerator Chi-Square Results 227
29.2 Summary Statistics on the Number of Runs for All Tests Using
      Quasi-Random Numbers 229
29.3 Results of All "Runs Tests" 230
29.4 Approximate Times Product P_i is in Department D_j 232